GOODS OF THE MIND, LLC

Competitive Mathematics Series
for
Gifted Students in Grades 3 and 4

PRACTICE OPERATIONS

Cleo Borac, M. Sc.
Silviu Borac, Ph. D.

This edition published in 2014 in the United States of America.

Editing and proofreading: David Borac, B.Mus.
Technical support: Andrei T. Borac, B.A., PBK

Send all inquiries to:

Goods of the Mind, LLC
1138 Grand Teton Dr.
Pacifica
CA, 94044

Competitive Mathematics Series for Gifted Students
Level II (Grades 3 and 4)
Practice Operations
2[ed] edition

Contents

FOREWORD

The goal of these booklets is to provide a problem solving training ground starting from the earliest years of a student's mathematical development.

In our experience, we have found that teaching how to solve problems should focus not only on finding correct answers but also on finding better solution strategies. While the correct answer to a problem can typically be obtained in several different ways, not all these ways are equally useful for learning how to solve problems.

The most basic strategy is *brute force.* For example, if a problem asks for the number of ways Lila and Dina can sit on a bench, it is easy to write down all the possibilities: Dina, Lila and Lila, Dina. We arrive at this solution by performing all the possible actions allowed by the problem, leaving nothing to the imagination. For this last reason, this approach is called brute force.

Obviously, if we had to figure out the number of ways 30 people could stand in a line, then brute force would not be as practical, as it would take a prohibitively long time to apply.

Using brute force to obtain the correct answer for a simpler problem is not necessarily a useful learning experience for solving a similar problem that is more complex. Moreover, solving problems in a quantitative manner, assuming that the student can transfer simple strategies to similar but more complex problems, is not an efficient way of learning problem solving.

From this simple example, we see that the goal of *practicing* problem solving is different from the goal of problem solving. While the goal of problem solving is to obtain a correct answer, the goal of practicing problem solving is to acquire the ability to develop strategies, generate ideas, and combine approaches that are powerful enough to solve the problem at hand as well as future similar problems.

While brute force is not a useless strategy, it is not a key that opens every door. Nevertheless, there are problems where brute force can be a useful tool. For instance, brute force can be used as a first step in solving a complex problem: a smaller scale example can be approached using brute force to help the problem solver understand the mechanics of the problem and generate ideas for solving the larger case.

All too often, we encounter students who can quickly solve simple problems by applying brute force and who become frustrated when the solving methods they have been employing successfully for years become inefficient once problems increase in complexity. Often, neither the student nor the parent has a clear understanding of why the student has stagnated at a certain level. When the only arrows in the quiver are guess-and-check and brute force, the ability to take down larger game is limited.

Our series of books aims to address this tendency to continue on the beaten path - which usually generates so much praise for the gifted student in the early years of schooling - by offering a challenging set of questions meant to build up an understanding of the problem solving process. Solving problems should never be easy! To be useful, to represent actual training, problem solving should be challenging. There should always be a sense of difficulty, otherwise there is no elation upon finding the solution.

Indeed, practicing problem solving is important and useful only as a means of learning how to develop better strategies. We must constantly learn and invent new strategies while questioning the limitations of the strategies we are using. Obtaining the correct answer is only the natural outcome of having applied a strategy that worked for a particular problem in the time available to solve it. Obtaining the wrong answer is not necessarily a bad outcome; it provides insight into the fallacies of the method used or into the errors of execution that may have occured. As long as students manifest an interest in figuring out strategies, the process of problem solving should be rewarding in itself.

Sitting and thinking in a focused manner is difficult to train, particularly since the modern lifestyle is not conducive to adopting open-ended activities. This is why we would like to encourage parents to pull back from a quantitative approach to mathematical education based on repetition, number of completed pages, and the number of correct answers. Instead, open up the time boundaries that are dedicated to math, adopt math as a game played in the family, initiate a math dialogue, and let the student take his or her time to think up clever solutions.

Figuring out strategies is much more of a game than the mechanical repetition of stepwise problem solving recipes that textbooks so profusely provide, in order to "make math easy." Mathematics is not meant to be easy; it is meant to be interesting.

Solving a problem in different ways is a good way of comparing the merits of each method - another reason for not making the correct answer the primary goal of the activity. Which method is more labor intensive, takes more time or is more prone to execution errors? These are questions that must be part of the problem solving process.

In the end, it is not the quantity of problems solved, the level of theory absorbed, or the number of solutions offered in ready-made form by so many courses and camps, but the willingness to ask questions, understand and explore limitations, and derive new information from scratch, that are the cornerstones of a sound training for problem solvers.

These booklets are not a complete guide to the problem solving universe, but they are meant to help parents and educators work in the direction that, aside from being the most efficient, is the more interesting and rewarding one.

The series is designed for mathematically gifted students. Each book addresses an age range as some students will be ready for this content earlier, others later. If a topic seems too difficult, simply try it again in a couple of months.

FRACTIONS

What is a fraction?

A quantity was divided into 13 equal parts

11 of these 13 parts are symbolized by the fraction

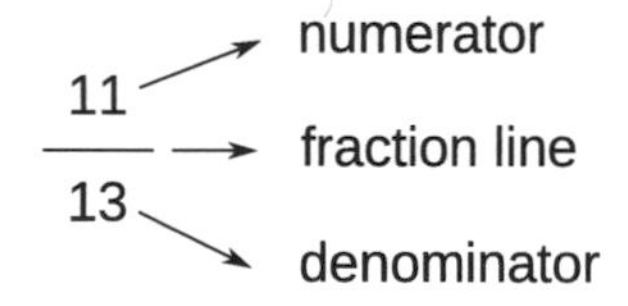

- The denominator of a fraction cannot be zero.
- An *improper fraction* has a numerator larger than the denominator and is larger than 1 as a result.
- A *proper fraction* has a numerator smaller than the denominator and is smaller than 1 as a result.
- Fractions with a denominator of 1 are integers.
- If the numerator and the denominator have *common factors*, the fraction can be *simplified* by dividing both the numerator and the denominator by a *common factor*.
- A fraction that cannot be simplified is an *irreducible fraction* and is said to be *in its lowest terms* (also *in its simplest form*).
- There are an infinite number of fractions that are equivalent to each irreducible fraction:

$$\frac{2}{3}, \frac{4}{6}, \frac{6}{9}, \ldots$$

To find out if the fraction $\frac{3}{11}$ is larger than the fraction $\frac{4}{13}$:

- Cross multiply:

$$3 \times 13 \qquad 11 \times 4$$

- Compare the products and find out which one is larger:

$$3 \times 13 < 11 \times 4$$

- The larger fraction is the one whose numerator is part of the larger product:

$$\frac{3}{11} < \frac{4}{13}$$

The fractions must be positive for this to work!

Example:

$$\frac{5}{8} \times \frac{9}{15}$$

$$8 \times 9 \qquad (5) \times 15$$

$$72 < (75)$$

$$\frac{(5)}{8} > \frac{9}{15}$$

Practice One

Do not use a calculator for any of the problems!

Exercise 1

Circle the fractions that are equivalent to $\frac{3}{5}$:

$\frac{69}{115}$ $\frac{18}{30}$ $\frac{9}{15}$ $\frac{9}{10}$ $\frac{18}{25}$ $\frac{12}{20}$ $\frac{6}{10}$

Exercise 2

Place the appropriate comparison operator ($=, >$, or $<$) between the fractions in each pair:

$\frac{3}{5}$ $\frac{3}{6}$	$\frac{11}{14}$ $\frac{143}{182}$	$\frac{5}{51}$ $\frac{5}{50}$	$\frac{135}{195}$ $\frac{9}{13}$
$\frac{18}{17}$ $\frac{18}{7}$	$\frac{187}{143}$ $\frac{17}{13}$	$\frac{5}{9}$ $\frac{15}{26}$	$\frac{8}{16}$ $\frac{51}{100}$
$\frac{2}{11}$ $\frac{242}{1331}$	$\frac{4}{31}$ $\frac{12}{96}$	$\frac{15}{13}$ $\frac{75}{65}$	$\frac{97}{98}$ $\frac{98}{99}$

Exercise 3

For each of the following, find answers that are irreducible fractions:

1. On the bakery's shelves, Max has 21 loaves, of which 14 are boules. What fraction of the loaves on display are boules?
2. Stephan has 50 tennis balls in his bag. 15 of them are Wilson balls and the rest are Dunlap balls. What fraction of the balls are Dunlap brand?
3. In Dina's class, there are 14 girls and 18 boys. What fraction of the students are boys?
4. The Roman general Flavius Aetius commanded around $50,000$ soldiers in the Catalaunian Plains battle, while Attila the Hun commanded around $30,000$ fighters. Approximately what fraction of the total number of troops were under the Roman flag?
5. Alfonso packaged an order of 40 lbs of tomatoes, 50 lbs of potatoes, and 30 lbs of onions for a restaurant. What fraction of the total weight did the tomatoes represent?

Exercise 4

If the positive fraction $\frac{k}{5}$ is proper, how many possible values are there for k?

Exercise 5

If the positive fraction $\dfrac{12}{12-k}$ is irreducible, how many positive values are there for k?

Exercise 6

How many improper fractions $\dfrac{t}{10}$ are there if $0 < t < 20$?

Exercise 7

How many irreducible proper fractions are there of the form $\dfrac{m}{10}$ if m is a positive integer?

Exercise 8

Compare the fractions, knowing that the ♣ is a positive integer:

$$\frac{\clubsuit + 1}{\clubsuit} \qquad \frac{\clubsuit}{\clubsuit + 1}$$

Exercise 9

Find all the possible positive integer values of k:

$$\frac{3}{5} > \frac{k}{7}$$

Exercise 10

Find all the possible values of the digit replaced by A such that the 2-digit number $1A$ satisfies:

$$\frac{1A}{15} < \frac{11}{13}$$

Exercise 11

Find all the possible values of the non-zero digit B such that:

$$\frac{7}{31} < \frac{17}{B2}$$

Exercise 12

Find the sum of the digits a and b knowing that:

$$\frac{3}{a3} = \frac{2a}{ba}$$

Exercise 13

Write each improper fraction as the sum of an integer and a proper fraction:

$\frac{11}{5}$	$\frac{111}{47}$	$\frac{55}{7}$	$\frac{195}{11}$
$\frac{39}{5}$	$\frac{k+1}{k}$	$\frac{k+20}{k+19}$	$\frac{m+5}{m+1}$

Exercise 14

Fill in the blanks:

1. One third of one third is one ....
2. One half of one third is one ....
3. Two thirds of one half is one ....
4. Three fifths of five thirds is one ....

Exercise 15

Find the value of the star:

$\frac{1}{3}$ of ☆ is ◇

$\frac{1}{4}$ of ◇ is 2

Exercise 16

Knowing that a and b are different digits, find the largest irreducible proper fraction of the form:

$$\frac{4ab}{501}$$

Exercise 17

Knowing that a and b are different digits, find the largest and the smallest proper positive fractions of the form:

$$\frac{ab5}{5ab}$$

Exercise 18

Write the improper fractions below in the form $k + \frac{m}{n}$, where k is an integer and $\frac{m}{n}$ is an irreducible proper fraction:

$\frac{11}{4} =$ $\qquad$ $\frac{23}{7} =$

$\frac{211}{42} =$ $\qquad$ $\frac{902}{17} =$

k is the largest integer smaller than the fraction or equal to it.

Exercise 19

Write the improper fractions below in the form $p - \frac{m}{n}$, where p is an integer and $\frac{m}{n}$ is an irreducible proper fraction:

$\frac{11}{4} =$ $\frac{23}{7} =$

$\frac{211}{42} =$ $\frac{902}{17} =$

p is the smallest integer larger than the fraction or equal to it.

Exercise 20

Frame each fraction between the largest integer smaller than it and the smallest integer larger than it:

$$< \tfrac{113}{14} <$$

$$< \tfrac{2}{7} <$$

$$< \tfrac{217}{19} <$$

$$< \tfrac{511}{12} <$$

The two integers must differ by 1.

Exercise 21

Find the positive integer x in each of the following:

1. $x + \frac{3}{5} = \frac{18}{5}$
2. $\frac{15}{19} + x = \frac{72}{19}$
3. $\frac{111}{34} = \frac{9}{34} + x$

Exercise 22

Find the positive integer x in each of the following:

1. $x - \frac{2}{5} = \frac{18}{5}$

2. $x - \frac{5}{9} = \frac{4}{9}$

3. $x - \frac{7}{11} = \frac{26}{11}$

Expressions

Processing expressions is often a creative and interesting task.

The goal of working with expressions should not simply be to find the correct answer, but to train solid skills for manipulating numeric expressions both on paper and mentally. Students should focus on:

- finding the simplest ways to produce partial answers
- adopting strategies that minimize error.

Here is a simple example:

$$9873 \times 4 \div 12$$

The student who multiplies 9873 by 4 and divides the result by 12, thus following the order of operations ad literam, is worse off - in terms of amount of work and probability of error - than the student who realizes that it is sufficient to divide 9873 by 3.

Similarly:

$$897614 + 1889453 - 897613$$

should not be computed by adding first and then subtracting. It is much easier to subtract 897613 from 897614 and add the result to 1889453.

When reducing expressions, it is important to *keep the numbers as small as possible at all times.*

- Addition and subtraction have *the same priority.* They can be performed in the order that is more advantageous:

$$\begin{aligned} 101 + 201 - 99 - 198 &= 101 - 99 + 201 - 198 \\ &= 2 + 3 \\ &= 5 \end{aligned}$$

- Multiplication and division have *the same priority.* They can be performed in the order that is more advantageous:

$$983 \div 7 \times 14 = 983 \times 14 \div 7 = 983 \times 2 = 1966$$

  This is particularly useful here since 983 is not divisible by 7.

- When processing expressions that contain both multiplications and divisions, factoring into primes and simplifying common factors first may save time and improve accuracy:

$$\begin{aligned} \frac{39 \times 76}{26 \times 38} &= \frac{13 \times 3 \times 2 \times 2 \times 19}{13 \times 2 \times 19 \times 2} \\ &= \frac{\cancel{13} \times 3 \times \cancel{2} \times \cancel{2} \times \cancel{19}}{\cancel{13} \times \cancel{2} \times \cancel{19} \times \cancel{2}} \\ &= 3 \end{aligned}$$

- Rearranging terms and factors instead of strictly following P E M D A S is a useful technique. In the following example, do not perform the operation in the parentheses first! Rather, rearrange the operations, factor the numbers into primes, and simplify:

$$\begin{aligned} 7 \times 76 \div (19 \times 42) &= \frac{7 \times 76}{19 \times 42} \\ &= \frac{\cancel{7} \times \cancel{2} \times \cancel{19}}{\cancel{19} \times \cancel{2} \times 3 \times \cancel{7}} = \frac{1}{3} \end{aligned}$$

Students must develop the ability to use the distributive property of multiplication/division over addition/subtraction in two ways:

- expanding: $9 \times (11 + 10) = 9 \times 11 + 9 \times 10 = 99 + 90 = 189$
- factoring: $9 \times 11 + 9 \times 10 = 9 \times (11 + 10) = 9 \times 21 = 189$

Example:

$$\begin{aligned}
550 \div (16 \times 5 - 25) &= 500 \div (5 \times 16 - 5 \times 5) \\
&= 550 \div (5 \times (16 - 5)) \\
&= 550 \div (5 \times 11) \\
&= \frac{\cancel{5} \times \cancel{11} \times 10}{\cancel{5} \times \cancel{11}} \\
&= 10
\end{aligned}$$

Observations on how to use flexibility in applying PEMDAS:

- In line 1, we did not execute the parentheses first. Instead, we identified a factor common to both terms in the subtraction.
- In line 2, we did not execute the parentheses. Instead, we used factoring and actually *increased* the number of parentheses. This allowed us to work with smaller numbers.
- In line 4, instead of multiplying, we wrote the expression in fraction form and factored the numbers to help simplify factors.
- In line 4, we did not factor 10 into primes since it was not necessary to do so.

Conclusion: *Predicting* partial results is the cornerstone of making good decisions when optimizing the order of operations. Developing the ability to predict future results and their impact on computations is the study goal for this chapter.

Practice Two

> Do not use a calculator for any of the problems!

Exercise 1

Find an efficient way to compute the value of the expression:

$$100 \div (19 \times 2 + 31 \times 2) =$$

Exercise 2

Find an efficient way to compute the value of the expression:

$$(202020 \div 160 \times 24 - 10101) \div (370 - 111) =$$

Exercise 3

In each of the following, factor before multiplying and apply the commutative and associative properties to make the computation easier.

1. $1125 \times 16 =$

2. $350 \times 28 =$

3. $56 \times 125 =$

Exercise 4

Compute efficiently:

$$(7 \times (226 \times 3 + 113) \div 49 - 26) \div 29 - 3 =$$

Exercise 5

Compute efficiently:

$$404 \times 17 \div 1212 \times (319 - 145) + 34 =$$

Exercise 6

Compute efficiently:

$$(333 + 555) \div 4 \div (3 \times 18 + 57) \times (1020 \div 85 \div 24) =$$

Exercise 7

Which of the following expressions are equivalent to:

$$180 \times 535 \div (84 \times 1125)$$

Check all that apply.

(A) $180 \times 535 \div 84 \div 1125$

(B) $180 \times 535 \div 84 \times 1125$

(C) $\dfrac{180 \times 535}{84 \times 1125}$

(D) $\dfrac{180 \times 535}{84 \div 1125}$

Exercise 8

Which of the following expressions are equivalent to:

$$a \div b \times c \div d \times m \div n$$

Check all that apply.

(A) $(a \times c \times m) \div (b \times d \times n)$

(B) $(a \times c \times m) \div (b \div d \div n)$

(C) $(a \times c \times m) \times (b \div d \div n)$

(D) $\dfrac{a \times c \times m}{b \times d \times n}$

Exercise 9

Which of the following computations have errors? Check all that apply.

(A) $\dfrac{\cancel{31} \times 13 + 19}{\cancel{31} \times 19 + 13} = \dfrac{13 + 19}{19 + 13} = 1$

(B) $\dfrac{\cancel{31} \times 13}{\cancel{31} \times 19 + 1} = \dfrac{13}{20}$

(C) $\dfrac{1+1+1+1+1}{2+2+2+2+2} = \dfrac{1}{2}$

(D) $99 \times (1000 - 100) = 99 \times 1000 - 99 \times 100$

Exercise 10

N is a positive integer. Which of the following will have an integer result? Check all that apply.

(A) Multiplying N by 3 and dividing the result by 9.

(B) Dividing N by 11 and multiplying the result by 132.

(C) Subtracting 5 from N, dividing the result by 5, and multiplying by 10.

(D) Subtracting 1 from N, multiplying the result by 5, and dividing by 10.

Exercise 11

If N is a positive integer, which of the following expressions are even?

(A) $N \times (N - 1)$

(B) $N \times (N + 1)$

(C) $N \times N - 1$

(D) $2 \times N + 1$

Exercise 12

Find the largest 2-digit number with different digits that produces a quotient of 13 when divided by 7.

Exercise 13

Roman numeral time! Dina and Lila have received this worksheet. Even if they share the work, they still need help from you!

1. CXXX − XCIX =
2. XCII + CXVII =
3. LXV + XLIX =
4. CCCLIX + LVII =
5. CCCXLI − LXIX =
6. DIII + CMXX =
7. CD − CC =
8. DXIX − CCCLXII =
9. CDXI + CMXI =
10. CDLV − CCCXCIX =

Exercise 14

Amira had to convert the following Arabic numerals to Roman numerals. Try to do this as well and then check the solutions to see Amira's answers.

1005, 1500, 1900, 990, 1998, 2003, 58, 412, 598

Exercise 15

Group efficiently to calculate:

$$2 + 4 + 6 + 11 + 13 + 15 + 25 + 27 + 29 + 44 + 46 + 48 =$$

Exercise 16

Replace the symbols with the numbers 2, 3, 4, and 6 so that the operations are correct:

$$((\clubsuit + \spadesuit) \times \diamondsuit + \heartsuit) \div 7 = 4$$

Exercise 17

Replace the symbols with digits from 2 to 9 to obtain the largest possible result. Different symbols represent different digits.

$$\clubsuit \div 2 + \diamondsuit \times \heartsuit =$$

Exercise 18

Find the largest 3-digit number mnp with the digits m, n, and p that satisfies the cryptarithm:

$$mnp \times p = 936$$

Operations with Repdigits

A *repdigit* is a number in which all the digits are identical. Examples of repdigits:

$$5555555$$
$$33333$$
$$\underbrace{77\cdots7}_{1980\text{ times}}$$

Example 1: Addition of 2 repdigits with carryover.

$$\begin{array}{r} 88\cdots888 \\ + 66\cdots666 \\ \hline 155\cdots554 \end{array}$$

(carries: +1 +1 … +1 +1)

Example 2: Addition of 3 repdigits with carryover.

$$\begin{array}{r} 88\cdots888 \\ 66\cdots666 \\ + 55\cdots555 \\ \hline 200\cdots009 \end{array}$$

(carries: +2 +2 … +2 +1)

Example 1: Subtraction of 2 repdigits with borrowing.

$$\begin{array}{r} 88\cdots888 \\ 9\cdots999 \\ \hline 78\cdots889 \end{array} -$$

Example 2: Multiplication of 2 repdigits.

$$\begin{array}{r} 88\cdots888 \\ 99 \\ \hline 79\cdots9992 \\ 799\cdots992 \\ \hline 879\cdots9912 \end{array} \times$$

Using the fact that $99 = 100 - 1$, another solution is possible:

$$888\cdots88 \times 99 = 888\cdots88 \times (100 - 1)$$

$$\begin{array}{r} 8888\cdots800 \\ 88\cdots888 \\ \hline 8799\cdots912 \end{array} -$$

Practice Three

Do not use a calculator for any of the problems!

Exercise 1

How many digits of 1 are there in the number:

$$\underbrace{1010\cdots101}_{131\text{ digits}}$$

Exercise 2

How many digits of 4 are there in the result of each operation:

(A) $8888 + 5555 =$

(B) $8888888 + 5555555 =$

(C) $\underbrace{88\cdots8}_{100\text{ digits}} + \underbrace{55\cdots5}_{100\text{ digits}} =$

(D) $\underbrace{88\cdots8}_{51\text{ digits}} + \underbrace{55\cdots5}_{51\text{ digits}} =$

Exercise 3

How many digits of 3 are there in the result of the operation:

$$\underbrace{44\cdots44}_{44\text{ digits}}5 \times 3$$

Exercise 4

How many digits of 7 are there in the result of the operation:

$$\underbrace{88\cdots8}_{100 \text{ digits}} + \underbrace{88\cdots8}_{50 \text{ digits}} =$$

Exercise 5

How many digits of 1 are there in the result of the operation:

$$\underbrace{55\cdots55}_{100 \text{ digits}}6 \times 11$$

Exercise 6

How many digits of 3 and how many digits of 7 are there in the result of each operation:

(A) $77777 + 5555 =$

(B) $77777777 + 55555 =$

(C) $\underbrace{77\cdots7}_{20 \text{ digits}} + \underbrace{55\cdots5}_{10 \text{ digits}} =$

(D) $\underbrace{77\cdots7}_{77 \text{ digits}} + \underbrace{55\cdots5}_{55 \text{ digits}} =$

Exercise 7

How many digits of 3 are there in each difference?

(A) $222222 - 88888 =$

(B) $222222222 - 888888 =$

(C) $\underbrace{22\cdots2}_{20 \text{ digits}} - \underbrace{88\cdots8}_{19 \text{ digits}} =$

(D) $\underbrace{22\cdots2}_{222 \text{ digits}} - \underbrace{88\cdots8}_{88 \text{ digits}} =$

Exercise 8

How many digits of 7 are there in the result of the operation?

$$\underbrace{77\cdots7}_{777\text{ digits}}\underbrace{33\cdots3}_{333\text{ digits}}-\underbrace{55\cdots5}_{555\text{ digits}}=$$

Exercise 9

If k and m represent numbers of digits, what are the values of k and m?

$$\underbrace{88\cdots8}_{k\text{ digits}}+\underbrace{55\cdots5}_{m\text{ digits}}=\underbrace{88\cdots8}_{12\text{ digits}}9\underbrace{44\cdots4}_{12\text{ digits}}3$$

Exercise 10

If we divide a repdigit formed of digits of 8 by 7 and want the division to be exact, which of the following numbers could be the number of digits of the repdigit?

(A) 7

(B) 8

(C) 12

(D) it is not possible to produce an exact division

Identities, Equations and Inequalities

Identities are equalities that are always true. For example:

$$4 = 4$$

and

$$x = x$$

are both identities. The quantity x can have any value: the identity will still be true.

Equations are equalities that are true only for certain values of the unknown quantities. For example:

$$x + 4 = 9$$

is true only if $x = 5$.

Inequalities are comparisons between two numbers or expressions. Here are a few examples:

$$4 > 2$$

and

$$4 + n > 12$$

An inequality like the last one is true only if the unknown quantity has values within a certain range. In this specific case, n must be larger than 8.

Linear Equations

Simple equations can be solved by working backwards to find the unknown quantity.

$$2 \times x + 5 = 11$$

The result of the operations on the left side is 11.
Since the last operation was to add 5, subtract 5 from 11.
x was multiplied by 2 to get 6; divide 6 by 2 to get 3.

When we solve word problems, it is often more efficient to *set up an equation* if we want to find the value of some unknown amount. To set up an equation correctly, you must make sure you:

- comprehend the statement;
- select a quantity that remains unchanged throughout the problem: the distance covered by two vehicles, the time needed for two simultaneous motions to execute, the capacity of a vessel, etc.;
- compute the unchanged quantity in two different ways and set the results to be equal.

Identifying the unchanged (invariant) quantity is often a challenge. Look for the occurence of the word "is" in the statement of the problem. This word usually marks the position of the equal sign in the equation.

Diophantine equations are equations in which all the numbers involved are integers.

Often, Diophantine equations have several unknown quantities and cannot be solved by working backwards. The methods used to solve these equations are based on the properties of integers, such as:

- the uniqueness of the prime factorization of a number,
- the parity of numbers (even or odd),
- narrowing down the range of values for a number,
- the range of values digits can have (between 0 and 9),
- the properties of consecutive numbers.

Example 1 Here is a Diophantine equation solved by factoring:

$$a \times b = 15$$
$$a \times b = 3 \times 5$$

If a and b must be *positive integers*, the possible solutions are:

$$a = 3 \quad b = 5$$
$$a = 5 \quad b = 3$$
$$a = 15 \quad b = 1$$
$$a = 1 \quad b = 15$$

If a and b must be *integers*, then we must also take into account the negative solutions:

$$a = -3 \quad b = -5$$
$$a = -5 \quad b = -3$$
$$a = -15 \quad b = -1$$
$$a = -1 \quad b = -15$$

Example 2 Solving a Diophantine equation by using the properties of consecutive numbers.

If m is an integer between 15 and 20 and can be written as

$$m = (k - 1) \times k \times (k + 1)$$

where k is an integer, what are the possible values of m?

Since $k - 1$, k, and $k + 1$ are 3 consecutive integers, there must be at least one even number among them and at least one multiple of 3 among them. Therefore, m must be divisible by 6. The only multiple of 6 between 15 and 20 is 18.

Example 3 Solving a Diophantine equation by using parity and by narrowing down the range of an unknown quantity.

Lila gave 2 scoops of grain to each hen and 3 scoops of grain to each rooster in the chicken coop. If she distributed a total of 20 scoops, what is the largest possible number of roosters?

Since 20 is *even*, the number of scoops given to hens and the number of scoops given to roosters must be either both *even* or both *odd*. Since the hens receive 2 scoops each, the number of scoops given to hens can only be *even*. Hence, the number of scoops given to roosters must also be *even*. Since the roosters receive 3 scoops each, the number of scoops received by roosters must be a multiple of 3. Since the number of scoops is even, it must also be a multiple of 6.

In how many ways can we write 20 as a sum of two even numbers, where one of them is a multiple of 6?

$$\begin{aligned} 20 &= 6 + 14 &&\rightarrow \text{2 roosters and 7 hens} \\ 20 &= 12 + 8 &&\rightarrow \text{4 roosters and 4 hens} \\ 20 &= 18 + 2 &&\rightarrow \text{6 roosters and 1 hen} \end{aligned}$$

PRACTICE FOUR

> Do not use a calculator for any of the problems!

Exercise 1

Find all the pairs of integers (m, n) that satisfy the equality:

$$77 = (m + 4)(n - 4)$$

Exercise 2

Find all the integer(s) k that satisfy the equality:

$$(k - 5) \times (k + 3) = 33$$

Exercise 3

Find the value of x:

$$((((((x + 4) - 5) + 6) - 7) + 8) - 9) + 10 = 3$$

Exercise 4

Find the value of x:

$$x + 4 - (5 + 6 - (7 + 8 - (9 + 10))) = 4$$

Exercise 5

Which digit has been replaced by the ♣?

$$\underbrace{\heartsuit\heartsuit\cdots\heartsuit}\times 7 = 6\underbrace{\clubsuit\clubsuit\cdots\clubsuit}16$$

Exercise 6

Ali and Baba discovered an underground palace and set out to map all the treasure inside it. After carefully mapping the interior, they found that the palace had a vault filled with treasure chests. In each chest, they found either 7 gold bars or 11 silver bars. There were 90 bars in total. How many of them were gold bars?

Exercise 7

When Lila divides a number by 11, she gets the same result as when she subtracts 80 from it. What is the number?

Exercise 8

Dina helped organize a raffle at the farmer's market. She packaged apples in bags of 5 and quinces in bags of 4. In total, she packaged 111 apples and quinces. There were fewer quinces than apples but fewer apples than twice the quinces. How many bags of each type of fruit did she package?

Exercise 9

Which triangular number is also a 3-digit repdigit?

Exercise 10

Find the value of x:

$$\frac{5}{x}\times\frac{4}{5}\times\frac{3}{4}\times\frac{2}{3}=\frac{1}{4}$$

Exercise 11

Find the value of x and write it in the form $n + \frac{a}{b}$ where n is an integer and $\frac{a}{b}$ is a proper irreducible fraction.

$$\frac{1}{x} = \frac{2+1}{2+4} \times \frac{3+1}{3+4} \times \frac{4+1}{4+4} \times \frac{5+1}{5+4} =$$

Exercise 12

(A) Instead of adding 17 to a number, Dina subtracted 17. Which number should she add to her result to get the correct answer?

(B) Instead of subtracting 101 from a number, Lila added 101. Which number should she subtract from her result to get the correct answer?

(C) Instead of multiplying a number by 12, Amira divided it by 4. What should she do in order to obtain the correct answer?

(D) Instead of dividing a number by 4, Dina multiplied it by 16. What should she do in order to obtain the correct answer?

Exercise 13

Amira wanted to make a larger cube out of 12 small cubes. When she finished, she realized she had not used all the small cubes. Which of the following could be the number of small cubes left over? (check all that apply)

(A) 0

(B) 2

(C) 3

(D) 4

Exercise 14

Find A, B, and C, knowing that:

$$\begin{aligned} A+B+C &= 50 \\ B+B+C &= 38 \\ C+C+C &= 36 \end{aligned}$$

Exercise 15

Find the positive integers A, B, and C, knowing that:

$$\begin{aligned} A \times B \times C &= 210 \\ B \times B \times C &= 252 \\ C \times C \times C &= 343 \end{aligned}$$

Exercise 16

Find the positive integers A, B, and C, knowing that:

$$\begin{aligned} A+B &= 25 \\ B+C &= 38 \\ A+C &= 47 \end{aligned}$$

Measurement

Physical quantities are measured in *units*. A measurement shows how many times larger than the base unit a certain physical quantity is.

Numeric comparisons of physical quantities are meaningful only if the quantities are all measured using the same unit.

Time is measured in seconds. Other units of time are multiples or divisors of the second:

1 minute = 60 seconds

1 hour = 60 minutes = 3600 seconds

1 day = 24 hours

1 week = 7 days

1 year = 365 days (or 366)

1 second = 1000 milliseconds

Notice how weird it is that multiples of the second use a multiplier of 60 while divisors of the second use a multiplier of 10.

Capacity is measured in liters. Multiples of the liter follow a pattern common to many metric units:

1 **kilo**liter = 1000 liters

1 **hecto**liter = 100 liters

1 **deca**liter = 10 liters

1 liter

10 **deci**liters = 1 liter

100 **centi**liters = 1 liter

1000 **milli**liters = 1 liter

Rates are measured in some unit per unit of time. Examples:

speed - miles per hour or meters per second

flow - liters per minute, gallons per hour

bit rate - bits per second

RPM - rotations per minute

Surface areas are measured in square units of length:

1 square kilometer = 1000000 square meters

1 square hectometer = 10000 square meters

1 square decameter = 100 square meters

1 square meter

100 square decimeters = 1 square meter

10000 square centimeters = 1 square meter

1000000 square millimiters = 1 square meter

To measure the surface area of a rectangle, we must measure the length and the width using the same unit. If we multiply the two numbers we get the surface area of the rectangle expressed in that square unit.

Temperature is measured using degrees in some scale of measurement. The Fahrenheit scale is the most frequently used in daily life in North America. In Europe, the Celsius (or centigrade) scale is used. Scientists use yet another scale: the Kelvin scale.

Temperatures can be converted from Fahrenheit to Celsius and back using the following formulas:

$$F = C \times 1.8 + 32$$

$$C = (F - 32) \div 1.8$$

When you travel, it is useful to be familiar with both scales.

Practice Five

Do not use a calculator for any of the problems!

Exercise 1

A 40 cm stick is what fraction of 1 meter long?

Exercise 2

Fill in the missing values in the table:

Length	30 cm	? cm	75 cm	10 mm	? mm
Fraction of 1 meter		$\frac{2}{5}$			$\frac{3}{10}$

Exercise 3

Dina and Lila were throwing a summer party. They had ten 750 milliliter (mL) bottles of lemonade and twelve 1 liter (L) bottles of limeade. They filled glasses with 150 milliliters of either lemonade or limeade until all the bottles were empty. How many more glasses of limeade than of lemonade were there when they finished?

Exercise 4

5 hours and 20 minutes is what fraction of one day?

Exercise 5

A side of a square is 40 cm long. Dina divides the square into 7 squares of different sizes. What is the side length of the smallest square?

Exercise 6

The perimeter of each of the following figures is what fraction of the perimeter of a square with a side length of 20 units?

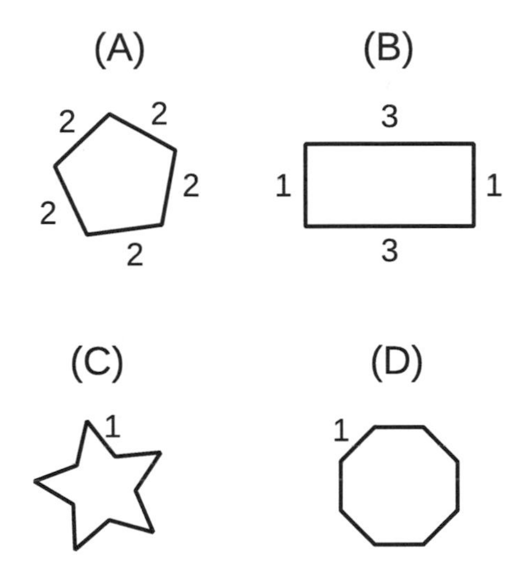

Exercise 7

Stephan, the tennis coach, must drive to a tournament. One of his students is also going, but in a separate vehicle. Stephan drives at a speed of 108 kilometers per hour, while the student drives at a speed of 30 meters per second. Who is driving faster?

(A) Stephan

(B) The student

(C) Both are driving at the same speed

Exercise 8

Amira decided to help her mother bake muffins. She measured 300 mL of milk in a measuring cup. Then, she put 6 squares of chocolate in the cup. After she added chocolate, the level in the measuring cup rose to 380 mL. What was the volume of the chocolate?

Exercise 9

For a school project, Dina and Lila had to measure a rectangular kitchen table. Dina measured the perimeter using a pencil as a unit and got a length of 34. Lila measured the perimeted using a bamboo skewer as a unit and got a length of 29. Which was longer, the pencil or the skewer?

Exercise 10

A square has a perimeter of 29 meters. Amira wanted to find a unit so that she would obtain a whole number of units when she measured the side of the square. Which of the following is a possible length of the unit?

(A) 8 cm

(B) 33 cm

(C) 58 cm

(D) 145 cm

Exercise 11

Ali wanted to get his donkey to return home with him from the Market. The donkey had a will of its own and did not want to go home without quenching his thirst, so they had to go to the Fountain first. The Fountain, however, is not on the street that leads from the Market to Ali's home, as the map in the figure shows:

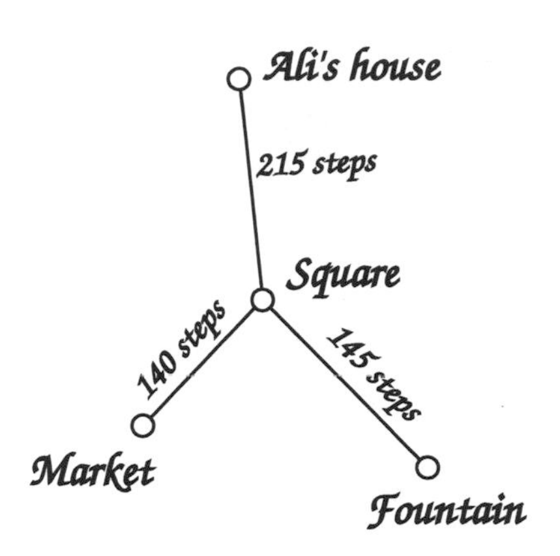

The distances shown on the map are measured in steps. How many steps did the donkey travel on its way from the Market to Ali's house?

Exercise 12

Dina's mother planted a square patch of raspberry bushes in the garden. Over the years, the raspberry plants grew outside the square. One day, Dina's mother noticed that, while the patch was still square, its perimeter had tripled. She asked Dina to find out how many times larger than before the area of the patch was. What did Dina find?

Miscellaneous Practice

Do not use a calculator for any of the problems!

Exercise 1

Write as an improper fraction:

$$\frac{1}{2} + \frac{1}{3} + \frac{1}{4} + \frac{1}{5}$$

Exercise 2

Write each of the fractions below as sums of two fractions with numerators that are consecutive:

$$\frac{11}{4} =$$

$$\frac{23}{7} =$$

$$\frac{211}{42} =$$

$$\frac{901}{17} =$$

Exercise 3

Solve the cryptarithm:

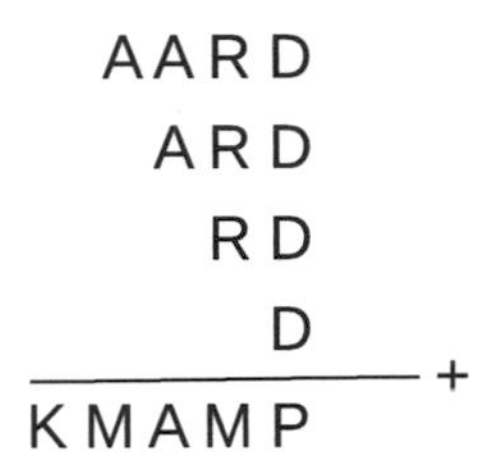

Exercise 4

Alfonso, the grocer, received a shipment of 95 unsorted avocadoes. He found that he was able to package the smaller avocadoes in bags of 9 each and the larger avocadoes in bags of 4 each without any avocadoes left over. How many small avocadoes were in the shipment? Select the solution that corresponds to the largest number of bags.

Exercise 5

If two circles have the same weight as three triangles and two triangles have the same weight as one square, which object should be removed from the balance to establish equilibrium?

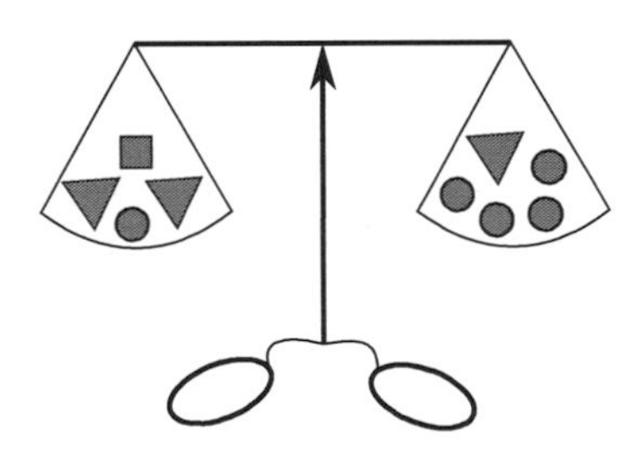

(A) a circle

(B) a triangle

(C) a square

Exercise 6

Compute efficiently:

1. $999 \times 1001 =$

2. $99 \times 1281 =$

3. $13 \times 125 =$

4. $ab \times 101 =$
   (ab is a 2-digit number with digits a and b)

5. $64 \times 75 =$

6. $9999 \times 9999 =$

Exercise 7

How many shaded squares are there?

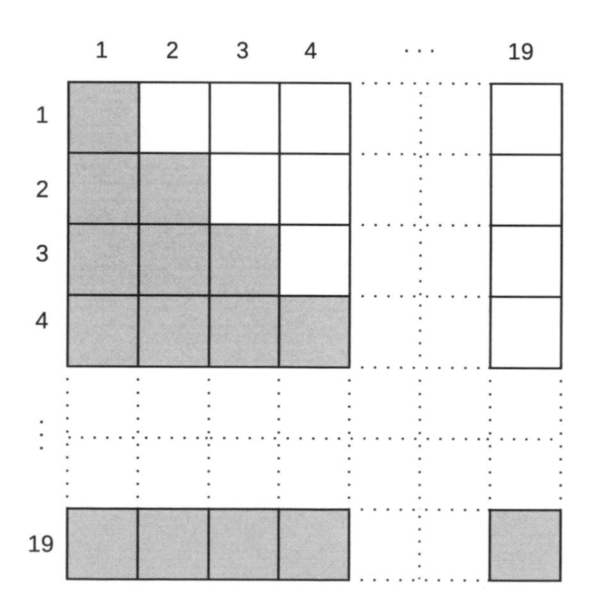

Exercise 8

Find x:

$$1 - (2 - (3 - (4 - (5 - (6 - (7 - (8 - x))))))) = 0$$

Exercise 9

Place parentheses so that the following is an identity (i.e. both sides represent the same value):

$$4 \times 4 - 4 \div 4 + 4 \times 4 = 4 \times 4 \times 4$$

Exercise 10

Place parentheses so that the result of the operations is an integer as large as possible:

$$5 \times 8 + 12 \div 3 + 5$$

Exercise 11

In the figure, each brick has a number that is the sum of the numbers on the two bricks below it. Which number is on the brick marked with an "X?"

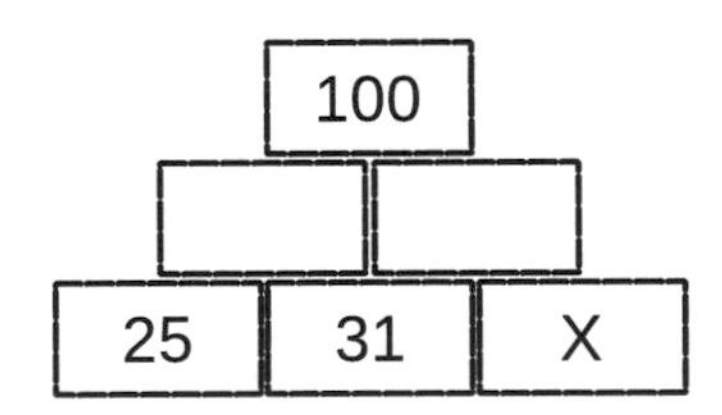

Exercise 12

In the figure, each brick has a number that is the product of the numbers on the two bricks below it. None of the numbers is equal to 1. Which number is on the brick marked with an "X?"

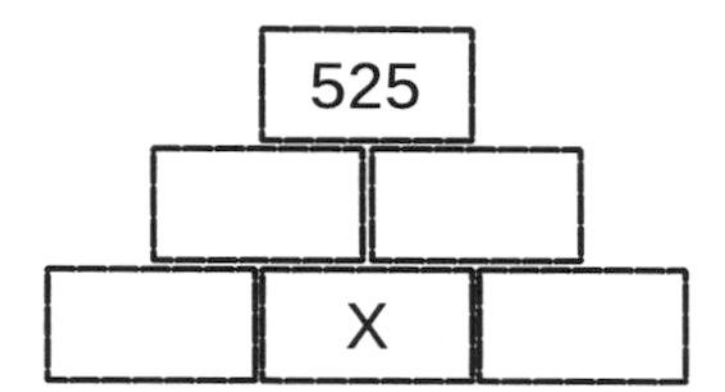

Exercise 13

Lila and Dina played a game of soccer while on opposing teams. The difference between the scores of the two teams was 4 times less than the total number of points scored. The sum of the total number of points scored and the difference between the scores was 40. If Lila was on the losing team, what score did her team have?

Exercise 14

How many 1 digits are there in the result of the operation:

$$\underbrace{10101\cdots10}_{100\text{ digits}} - \underbrace{9090\cdots90}_{99\text{ digits}}$$

Exercise 15

How many digits of 1 are there in the result of the operation:

$$\underbrace{10101\cdots10}_{100\text{ digits}} - \underbrace{9090\cdots90}_{49\text{ digits}}$$

Exercise 16

Find the consecutive positive integers k, l, and m:

$$(k-1)(l-1)(m-1) = 504$$

Exercise 17

Find a in each of the following:

1. $(a - 2.55 \div 3) \times 8 = \frac{6}{5}$
2. $450 - 171 \div a = 441$
3. $(41 \times a - 497) \div 149 = 14$
4. $4 \div (5 - 4 \div (3 + 3 \div a)) = 1$
5. $5 \times (a \times 1 \div a + a \times 0 - a \div 1 - 6 \times a \div 2) \div 11 = 95 \div 209$

Exercise 18

Amira has learned how to add, subtract, multiply, and divide multi-digit numbers. Because she is still little, it takes her some time to compute. She can perform an addition in 1 minute and a multiplication in 3 minutes. How quickly can Amira find the result of the operations:

$$3 \times 30 + 5 \times 30$$

Exercise 19

Lila is 4 years older than Amira. The product of their ages is 143. How old is Amira?

Exercise 20

Lila, Dina, and Amira use a shared piggy bank to save money for toys and jewelry. Before the winter holidays, they decided to use the money they had saved. Dina spent one quarter of the money on a bracelet, Lila spent one third of the remaining money on little pets, and Amira spent half of the remaining money on talking beads. Which of the following is true:

(A) the talking beads were the most expensive

(B) the bracelet was the most expensive

(C) the little pets were the most expensive

(D) they all cost the same

Solutions to Practice One

Exercise 1

Circle the fractions that are equivalent to $\frac{3}{5}$:

Solution 1

$$\frac{3}{5} = \frac{3}{5} \times \frac{6}{6} = \frac{18}{30}$$

$$\frac{3}{5} = \frac{3}{5} \times \frac{3}{3} = \frac{9}{15}$$

$$\frac{3}{5} = \frac{3}{5} \times \frac{23}{23} = \frac{69}{115}$$

$$\frac{3}{5} = \frac{3}{5} \times \frac{4}{4} = \frac{12}{20}$$

$$\frac{3}{5} = \frac{3}{5} \times \frac{2}{2} = \frac{6}{10}$$

None of the other choices are equivalent to $\frac{3}{5}$.

Exercise 2

Place the appropriate comparison operator ($=, >$, or $<$) between the fractions in each pair:

Solution 2

$\frac{3}{5} > \frac{3}{6}$	$\frac{11}{14} = \frac{143}{182}$	$\frac{5}{51} < \frac{5}{50}$	$\frac{135}{195} = \frac{9}{13}$
$\frac{18}{17} < \frac{18}{7}$	$\frac{187}{143} = \frac{17}{13}$	$\frac{5}{9} < \frac{15}{26}$	$\frac{8}{16} < \frac{51}{100}$
$\frac{2}{11} = \frac{242}{1331}$	$\frac{4}{31} > \frac{12}{96}$	$\frac{15}{13} = \frac{75}{65}$	$\frac{97}{98} > \frac{96}{98}$

Exercise 3

For each of the following, find answers that are irreducible fractions:

1. On the bakery's shelves, Max has 21 loaves of which 14 are boules. What fraction of the loaves on display are boules?
2. Stephan has 50 tennis balls in his bag. 15 of them are Wilson balls and the rest are Dunlap balls. What fraction of the balls are Dunlap brand?
3. In Dina's class, there are 14 girls and 18 boys. What fraction of the students are boys?
4. The Roman general Flavius Aetius commanded around 50,000 soldiers in the Catalaunian Plains battle, while Attila the Hun commanded around 30,000 fighters. Approximately what fraction of the total number of troops were under the Roman flag?
5. Alfonso packaged an order of 40 lbs of tomatoes, 50 lbs of potatoes, and 30 lbs of onions for a restaurant. What fraction of the total did the weight of the tomatoes represent?

Solution 3

1. $\frac{14}{21} = \mathbf{\frac{2}{3}}$
2. $\frac{50-15}{50} = \frac{35}{50} = \mathbf{\frac{7}{10}}$
3. $\frac{18}{14+18} = \frac{18}{32} = \mathbf{\frac{9}{16}}$
4. $\frac{50,000}{50,000+30,000} = \mathbf{\frac{5}{8}}$
5. $\frac{40}{40+50+30} = \frac{4}{12} = \mathbf{\frac{1}{3}}$

Exercise 4

If the positive fraction $\frac{k}{5}$ is proper, how many possible values are there for k?

Solution 4

In a fraction, both the numerator and the denominator must be integers. A positive proper fraction is smaller than 1 and greater than 0. The possible values for the numerator are: 1, 2, 3, and 4. There are 4 possible values.

Exercise 5

If the positive fraction $\frac{12}{12-k}$ is irreducible, how many positive values are there for k?

Solution 5

Notice that k cannot be larger than 11 since the fraction has to be positive. This requirement limits the range of values for k from 0 to 11. For the fraction to be irreducible, 12 and $12-k$ must not have any common divisors.

The divisors of 12 are: 1, 2, 3, 4, 6, and 12. $12-k$ can only be 5, 7, or 11. There are 3 possible values for k: 7, 5, and 1.

Exercise 6

How many improper fractions $\frac{t}{10}$ are there if $0 < t < 20$?

Solution 6

The integer t can range from 11 to 19: a total of 9 different values. Therefore, there are 9 fractions that satisfy the conditions.

Exercise 7

How many irreducible proper fractions are there of the form $\frac{m}{10}$ if m is a positive integer?

Solution 7

m must be smaller than 10 for the fraction to be proper. For the fraction to be irreducible, m must not have factors of 2 or 5. The possible fractions are:

$$\frac{1}{10}, \frac{3}{10}, \frac{7}{10}, \frac{9}{10}$$

There are 4 fractions in total.

Exercise 8

Compare the fractions, knowing that the $\clubsuit$ is a positive integer:

$$\frac{\clubsuit + 1}{\clubsuit} \qquad \frac{\clubsuit}{\clubsuit + 1}$$

Solution 8

The fraction on the left is improper while the fraction on the right is proper. Therefore:

$$\frac{\clubsuit + 1}{\clubsuit} > \frac{\clubsuit}{\clubsuit + 1}$$

Exercise 9

Find all the possible positive integer values of k:

$$\frac{3}{5} > \frac{k}{7}$$

Solution 9

$$\begin{aligned} \frac{3}{5} &> \frac{k}{7} \\ 3 \times 7 &> k \times 5 \\ 21 &> 5 \times k \end{aligned}$$

k can only be 1, 2, 3, or 4.

Exercise 10

Find all the possible values of the digit replaced by A such that the 2-digit number $1A$ satisfies:

$$\frac{1A}{15} < \frac{11}{13}$$

Solution 10

We can multiply an inequality by a positive number without changing it. Multiply both sides by 13×15:

$$\begin{aligned} \frac{1A}{15} &< \frac{11}{13} \\ 1A \times 13 &< 11 \times 15 \\ (10 + A) \times 13 &< 165 \\ 130 + 13 \times A &< 165 \\ 13 \times A &< 35 \end{aligned}$$

A can only be 1 or 2.

Exercise 11

Find all the possible values of the non-zero digit B such that:

$$\frac{7}{31} < \frac{17}{B2}$$

Solution 11

Since both denominators are guaranteed to be positive, multiply the inequality by $31 \times B2$:

$$\begin{aligned} \frac{7}{31} &< \frac{17}{B2} \\ 7 \times B2 &< 17 \times 31 \\ 7 \times (10 \times B + 2) &< 527 \\ 70 \times B + 14 &< 527 \\ 70 \times B &< 513 \end{aligned}$$

Since:

$$70 \times 7 < 513 < 70 \times 8$$

there are 7 possible values for the digit B: 1, 2, 3, 4, 5, 6, and 7.

Exercise 12

Find the sum of the digits a and b knowing that:

$$\frac{3}{a3} = \frac{2a}{ba}$$

Solution 12

For the fractions to be equivalent, the fraction on the left must be multiplied by a fraction equivalent to 1 that makes the numerator a 2-digit multiple of 3 that is between 20 and 30. There are only 3 such multiples: 21, 24, and 27. They represent multiplications by:

$$\frac{7}{7}, \frac{8}{8}, \frac{9}{9}$$

Of these, only 21 produces a solution:

$$\frac{3}{13} \times \frac{7}{7} = \frac{21}{91}$$

$a + b = 1 + 9 = 10$

Exercise 13

Write each improper fraction as the sum of an integer and a proper fraction:

Solution 13

$\frac{11}{5} = 2 + \frac{1}{5}$	$\frac{111}{47} = 2 + \frac{17}{47}$	$\frac{55}{7} = 7 + \frac{6}{7}$	$\frac{195}{11} = 17 + \frac{8}{11}$
$\frac{39}{5} = 7 + \frac{4}{5}$	$\frac{k+1}{k} = 1 + \frac{1}{k}$	$\frac{k+20}{k+19} = 1 + \frac{1}{k+19}$	$\frac{m+5}{m+1} = 1 + \frac{4}{m+1}$

Solution 14

1. One third of one third is one ninth.
2. One half of one third is one sixth.
3. Two thirds of one half is one third.
4. Three fifths of five thirds is one.

Exercise 15

Find the value of the star:

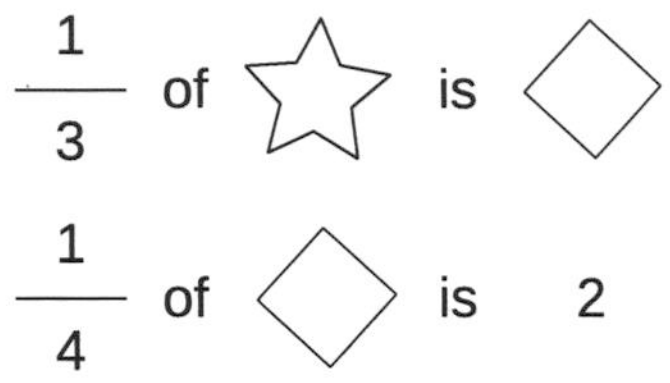

$$\frac{1}{3} \text{ of } ☆ \text{ is } ◇$$

$$\frac{1}{4} \text{ of } ◇ \text{ is } 2$$

Solution 15

Work backwards from the only quantity provided: 2. Since 2 is one quarter of the square, the square is 8. Since 8 is one third of the star, the star is 24.

Exercise 16

Knowing that a and b are different digits, find the largest irreducible proper fraction of the form:

$$\frac{4ab}{501}$$

Solution 16

Make the numerator the largest possible integer of the required form:

$$\frac{498}{501}$$

and notice that the fraction is reducible by 3:

$$\frac{2 \times 3 \times 83}{3 \times 167} = \frac{166}{167}$$

To make the fraction of the given form irreducible go to the next lower numerator:

$$\frac{497}{501}$$

Since $497 = 7 \times 71$, it is relatively prime (*coprime*) with 501 and the fraction is irreducible. Therefore, $a = 9$ and $b = 7$.

Exercise 17

Knowing that a and b are different non-zero digits, find the largest and the smallest proper positive fractions of the form:

$$\frac{ab5}{5ab}$$

Solution 17

For simplicity, denote the 2-digit number ab by n. Write the numerator and the denominator in exapanded form and compare to 1:

$$\frac{10 \times n + 5}{500 + n} < 1$$

Since $500 + n$ is positive, multiplying both sides by it will not change the inequality:

$$10 \times n + 5 < 500 + n$$

Subtract 5 and n from both sides:

$$9 \times n < 495$$

Since $495 \div 9 = 55$, we must have $n < 55$. With a and b different digits, the largest value of n is 54. The fraction:

$$\frac{545}{554}$$

is the largest proper fraction possible.

The smallest proper fraction is:

$$\frac{105}{510}$$

Exercise 18

Write the improper fractions below in the form $k + \frac{m}{n}$, where k is an integer and $\frac{m}{n}$ is an irreducible proper fraction:

Solution 18

$$\frac{11}{4} = 2 + \frac{3}{4} \qquad \frac{23}{7} = 3 + \frac{2}{7}$$

$$\frac{211}{42} = 5 + \frac{1}{42} \qquad \frac{902}{17} = 53 + \frac{1}{17}$$

Exercise 19

Write the improper fractions below in the form $p - \frac{m}{n}$, where p is an integer and $\frac{m}{n}$ is an irreducible proper fraction:

Solution 19

$$\frac{11}{4} = 3 - \frac{1}{4} \qquad \frac{23}{7} = 4 - \frac{5}{7}$$

$$\frac{211}{42} = 6 - \frac{41}{42} \qquad \frac{902}{17} = 54 - \frac{16}{17}$$

Exercise 20

Frame each fraction between the largest integer smaller than it and the smallest integer larger than it:

Solution 20

$$8 < \tfrac{113}{14} < 9$$

$$0 < \tfrac{2}{7} < 1$$

$$11 < \tfrac{217}{19} < 12$$

$$42 < \tfrac{511}{12} < 43$$

Exercise 21

Find the positive integer x in each of thc following:

1. $x + \frac{3}{5} = \frac{18}{5}$

2. $\frac{15}{19} + x = \frac{72}{19}$

3. $\frac{111}{34} = \frac{9}{34} + x$

Solution 21

1. $\mathbf{3} + \frac{3}{5} = \frac{18}{5}$

2. $\frac{15}{19} + \mathbf{3} = \frac{72}{19}$

3. $\frac{315}{34} = \frac{9}{34} + \mathbf{9}$

Exercise 22

Find the positive integer x in each of the following:

1. $x - \frac{2}{5} = \frac{18}{5}$

2. $x - \frac{5}{9} = \frac{4}{9}$

3. $x - \frac{7}{11} = \frac{26}{11}$

Solution 22

1. $\mathbf{4} - \frac{2}{5} = \frac{18}{5}$

2. $\mathbf{1} - \frac{5}{9} = \frac{4}{9}$

3. $\mathbf{3} - \frac{7}{11} = \frac{26}{11}$

Solutions to Practice Two

Exercise 1

Find an efficient way to compute the value of the expression:

$$100 \div (19 \times 2 + 31 \times 2) =$$

Solution 1

$$\begin{aligned} 100 \div (19 \times 2 + 31 \times 2) &= 100 \div (2 \times (19 + 31)) \\ &= 100 \div (2 \times 50) \\ &= 100 \div 100 \\ &= \mathbf{1} \end{aligned}$$

Exercise 2

Find an efficient way to compute the value of the expression:

$$(202020 \div 160 \times 24 - 10101) \div (370 - 111) =$$

Solution 2

$$\begin{aligned} & (202020 \div 160 \times 24 - 10101) \div (370 - 111) \\ = \; & \frac{\not{20} \times 10101 \times \not{8} \times 3}{\not{20} \times \not{8}} \div (37 \times 10 - 37 \times 3) \\ = \; & \frac{10101 \times 3}{37 \times (10 - 3)} \\ = \; & \frac{3 \times \not{7} \times 13 \times \not{37}}{\not{37} \times \not{7}} \\ = \; & \mathbf{39} \end{aligned}$$

Exercise 3

In each of the following, factor before multiplying and apply the commutative and associative properties to make the computation easier.

1. $1125 \times 16 =$

2. $350 \times 28 =$

3. $56 \times 125 =$

Solution 3

1. $1125 \times 16 =$

$$
\begin{aligned}
&= 3 \times 3 \times 5 \times 5 \times 5 \times 2 \times 2 \times 2 \times 2 \\
&= 3 \times 3 \times 2 \times 5 \times 2 \times 5 \times 2 \times 5 \times 2 \\
&= 9 \times 2 \times 10 \times 10 \times 10 \\
&= 18 \times 1000 \\
&= 18,000
\end{aligned}
$$

2. $350 \times 28 =$

$$
\begin{aligned}
&= 7 \times 5 \times 10 \times 4 \times 7 \\
&= 49 \times 2 \times 2 \times 5 \times 10 \\
&= 49 \times 2 \times 100 \\
&= 98 \times 100 \\
&= 9,800
\end{aligned}
$$

3. $56 \times 125 =$

$$\begin{aligned} &= 7 \times 8 \times 125 \\ &= 7 \times 2 \times 5 \times 2 \times 5 \times 2 \times 5 \\ &= 7 \times 10 \times 10 \times 10 \\ &= 7{,}000 \end{aligned}$$

Note that we did not necessarily factor everything down to primes, but just so that it is more convenient to multiply.

It is useful to retain that:

$$\begin{aligned} 2 \times 5 &= 10 \\ 4 \times 25 &= 100 \\ 8 \times 125 &= 1000 \end{aligned}$$

Exercise 4

Compute efficiently:

$$(7 \times (226 \times 3 + 113) \div 49 - 26) \div 29 - 3 =$$

Solution 4

$$\begin{aligned}
& (7 \times (226 \times 3 + 113) \div 49 - 26) \div 29 - 3 \\
= \ & (7 \times (113 \times 2 \times 3 + 113) \div 49 - 26) \div 29 - 3 \\
= \ & (7 \times 113 \times (6 + 1) \div 49 - 26) \div 29 - 3 \\
= \ & (7 \times 7 \times 113 \div 49 - 26) \div 29 - 3 \\
= \ & (49 \div 49 \times 113 - 26) \div 29 - 3 \\
= \ & (113 - 26) \div 29 - 3 \\
= \ & \frac{87}{29} - 3 \\
= \ & \frac{3 \times \cancel{29}}{\cancel{29}} - 3 \\
= \ & 3 - 3 \\
= \ & \mathbf{0}
\end{aligned}$$

Exercise 5

Compute efficiently:

$$404 \times 17 \div 1212 \times (319 - 145) + 34 =$$

Solution 5

$$
\begin{aligned}
& 404 \times 17 \div 1212 \times (319 - 145) + 34 \\
= & \frac{\cancel{101} \times \cancel{4} \times 17}{\cancel{101} \times \cancel{4} \times 3} \times (11 \times 29 - 5 \times 29) + 34 \\
= & \frac{17}{3} \times 29 \times (11 - 5) + 34 \\
= & \frac{17}{\cancel{3}} \times 29 \times \cancel{3} \times 2 + 34 \\
= & 17 \times 29 \times 2 + 34 \\
= & 34 \times 29 + 34 \\
= & 34 \times (29 + 1) \\
= & 34 \times 30 \\
= & \mathbf{1020}
\end{aligned}
$$

Exercise 6

Compute efficiently:

$$(333 + 555) \div 4 \div (3 \times 18 + 57) \times (1020 \div 85 \div 24) =$$

Solution 6

$$\begin{aligned}
& (333 + 555) \div 4 \div (3 \times 18 + 57) \times (1020 \div 85 \div 24) \\
= \; & 888 \div 4 \div (3 \times 18 + 3 \times 19) \times \frac{1020}{85 \times 24} \\
= \; & \frac{111 \times \not{4} \times 2}{\not{4} \times 3 \times (18 + 19)} \times \frac{\not{5} \times \not{2} \times \not{2} \times \not{3} \times \not{17}}{\not{5} \times \not{17} \times \not{2} \times \not{2} \times 2 \times \not{3}} \\
= \; & \frac{\not{3} \times \not{37} \times 2}{\not{3} \times \not{37}} \times \frac{1}{2} \\
= \; & \not{2} \times \frac{1}{\not{2}} \\
= \; & \mathbf{1}
\end{aligned}$$

Exercise 7

Which of the expressions A-D are equivalent to:

$$180 \times 535 \div (84 \times 1125)$$

Check all that apply.

Solution 7

Expressions (A) and (C) are equivalent to the given expression.

(A) $\mathbf{180 \times 535 \div 84 \div 1125}$

(B) $180 \times 535 \div 84 \times 1125$

(C) $\mathbf{\dfrac{180 \times 535}{84 \times 1125}}$

(D) $\dfrac{180 \times 535}{84 \div 1125}$

Exercise 8

Which of the following expressions are equivalent to:

$$a \div b \times c \div d \times m \div n$$

Check all that apply.

Solution 8

Expressions (A) and (D) are equivalent to the given expression.

(A) $(\mathbf{a} \times \mathbf{c} \times \mathbf{m}) \div (\mathbf{b} \times \mathbf{d} \times \mathbf{n})$

(B) $(a \times c \times m) \div (b \div d \div n)$

(C) $(a \times c \times m) \times (b \div d \div n)$

(D) $\dfrac{\mathbf{a} \times \mathbf{c} \times \mathbf{m}}{\mathbf{b} \times \mathbf{d} \times \mathbf{n}}$

Exercise 9

Which of the following computations have errors? Check all that apply.

Solution 9

The incorrect computations (A) and (B) have been bolded.

(A) $\dfrac{\cancel{\mathbf{31}} \times 13 + 19}{\cancel{\mathbf{31}} \times 19 + 13} = \dfrac{13 + 19}{19 + 13} = 1$

(B) $\dfrac{\cancel{\mathbf{31}} \times 13}{\cancel{\mathbf{31}} \times 19 + 1} = \dfrac{13}{20}$

(C) $\dfrac{1+1+1+1+1}{2+2+2+2+2} = \dfrac{5}{10} = \dfrac{1}{2}$

(D) $99 \times (1000 - 100) = 99 \times 1000 - 99 \times 100$

Exercise 10

N is a positive integer. Which of the following will have an integer result? Check all that apply.

Solution 10

(A) "Multiply N by 3 and divide the result by 9," is equivalent to "Divide N by 3." Since we do not know if N is a multiple of 3, we cannot be sure that the result is an integer.

(B) "Divide N by 11 and multiply the result by 132," is equivalent to "Multiply N by 12." ($132 = 11 \times 12$) The result of the operation is an integer for any value of N.

(C) "Subtract 5 from N, divide the result by 5, and multiply by 10," is equivalent to "Subtract 5 from N and multiply the result by 2." This operation always produces an integer, albeit a negative one if the number we start with is smaller than 5.

(D) "Subtract 1 from N, multiply the result by 5, and divide by 10," is equivalent to "Subtract 1 from N and divide the result by 2." Since we do not know if N is even or odd, we cannot tell if $N - 1$ is a multiple of 2 or not. If N is even, the result is not an integer.

Exercise 11

If N is a positive integer, which of the following expressions are even?

Solution 11

(A) $N \times (N - 1)$ is even for all N. If N is odd, then $N - 1$ is even and the product is even. If N is even, then the product is even.

(B) $N \times (N + 1)$ is even for all N. If N is odd, then $N + 1$ is even and the product is even. If N is even, then the product is even.

(C) $N \times N - 1$ does not have known parity. If N is even, then $N \times N$ is even and the expression is odd. If N is odd, then $N \times N$ is odd and the expression is even.

(D) $2 \times N + 1$ is odd for all N, since the product of N and 2 is always an even number. Adding 1 to it makes the expression odd for all N.

Exercise 12

Find the largest 2-digit number with different digits that produces a quotient of 13 when divided by 7.

Solution 12

The largest 2-digit multiple of 13 is $91 = 13 \times 7 + 0$. Numbers between 91 and 98 produce a quotient of 13, and some non-zero remainder, when divided by 7. The remainder can be an integer from 1 to 6. Since the largest remainder is 6, the largest 2-digit number that gives a quotient of 13 is:

$$13 \times 7 + 6 = 97$$

The answer is **97**.

Exercise 13

Roman numeral time! Dina and Lila have received this worksheet. Even if they share the work, they still need help from you!

Solution 13

1. CXXX − XCIX = XXXIII
2. XCII + CXVII = CCIX
3. LXV + XLIX = CXIV
4. CCCLIX + LVII = CDXVI
5. CCCXLI − LXIX = CCLXXII
6. DIII + CMXX = MCDXXIII
7. CD − CC = CC
8. DXIX − CCCLXII = CLVII
9. CDXI + CMXI = MCCCXXII
10. CDLV − CCCXCIX = LVI

Exercise 14

Amira had to convert the following Arabic numerals to Roman numerals. Try to do this as well and then check the solutions to see Amira's answers.

1005, 1500, 1900, 990, 1998, 2003, 58, 412, 598

Solution 14

$$
\begin{aligned}
1005 &= \mathrm{MV} \\
1500 &= \mathrm{MD} \\
1900 &= \mathrm{MCM} \\
990 &= \mathrm{CMXC} \\
1998 &= \mathrm{MCMXCVIII} \\
2003 &= \mathrm{MMIII} \\
58 &+ \mathrm{LVIII} \\
412 &= \mathrm{CDXII} \\
598 &= \mathrm{DXCVIII}
\end{aligned}
$$

Exercise 15

Group efficiently to calculate:

$$2 + 4 + 6 + 11 + 13 + 15 + 25 + 27 + 29 + 44 + 46 + 48 =$$

Solution 15

Notice that there are groups of consecutive even and odd terms. A grouping that makes the computation easier is:

$$\begin{aligned}
& (2 + 48) + (4 + 46) + (6 + 44) + (11 + 29) + (13 + 27) + (15 + 25) \\
= \;& 50 + 50 + 50 + 40 + 40 + 40 \\
= \;& (50 + 40) \times 3 \\
= \;& 90 \times 3 \\
= \;& 270
\end{aligned}$$

Exercise 16

Replace the symbols with the numbers 2, 3, 4, and 6 so that the operations are correct:

$$((\clubsuit + \spadesuit) \times \diamondsuit + \heartsuit) \div 7 = 4$$

Solution 16

$$((2 + 6) \times 3 + 4) \div 7 = 4$$

Exercise 17

Replace the symbols with digits from 2 to 9 so as to obtain the largest possible integer result. Different symbols represent different digits.

$$\clubsuit \div 2 + \diamondsuit \times \heartsuit =$$

Solution 17

The ♣ must be an even digit. If we choose it to be the largest even digit we will not get the largest possible result. We are better off using the largest even digit as one of the factors of the next term:

$$6 \div 2 + 9 \times 8 = 3 + 72 = \mathbf{75}$$

Exercise 18

Find the largest 3-digit number mnp with the digits m, n, and p that satisfies the cryptarithm:

$$mnp \times p = 936$$

Solution 18

For the result to end in 6, p must be either 6 or 4.

If the last digit is 4, the solution is easy to find: $234 \times 4 = 936$.

By comparison, if we use a last digit of 6, the operation becomes: $156 \times 6 = 936$ and produces, as anticipated, a smaller number mnp.

Solutions to Practice Three

Exercise 1

How many digits of 1 are there in the number:

$$\underbrace{1010\cdots101}_{131\text{ digits}}$$

Solution 1

Since there are 131 digits in total, there are 65 pairs of 10 that are followed by a single digit of 1. There are $65+1=66$ digits of 1 and 65 digits of 0.

Exercise 2

How many digits of 4 are there in the result of each operation:

Solution 2

(A) $8888+5555=14443$

There are 3 digits of 4.

(B) $8888888+5555555=14444443$

There are 6 digits of 4.

(C) $\underbrace{88\cdots8}_{100\text{ digits}}+\underbrace{55\cdots5}_{100\text{ digits}}=1\underbrace{44\cdots4}_{99\text{ digits}}3$

There are 99 digits of 4.

(D) $\underbrace{88\cdots8}_{51\text{ digits}}+\underbrace{55\cdots5}_{51\text{ digits}}=1\underbrace{44\cdots4}_{50\text{ digits}}3$

There are 50 digits of 4.

Exercise 3

How many digits of 3 are there in the result of the operation:

$$\underbrace{44\cdots44}_{44\text{ digits}}5\times 3$$

Solution 3

The result of the operation has 46 digits:

$$\underbrace{44\cdots44}_{44\text{ digits}}5\times 3 = 1\underbrace{33\cdots33}_{44\text{ digits}}5$$

44 of them are digits of 3.

Exercise 4

How many digits of 7 are there in the result of the operation:

$$\underbrace{88\cdots8}_{100\text{ digits}}+\underbrace{88\cdots8}_{50\text{ digits}}=$$

Solution 4

There are 49 digits of 7 in the sum:

$$\begin{array}{r} \overbrace{888\cdots8}^{50\text{ digits}}\overbrace{88\cdots888}^{50\text{ digits}} \\ 88\cdots888 \\ \hline \underbrace{999\cdots9}_{50\text{ digits}}\underbrace{77\cdots77}_{49\text{ digits}}6 \end{array} +$$

Exercise 5

How many digits of 1 are there in the result of the operation:

$$\underbrace{55\cdots55}_{100\text{ digits}}6 \times 11$$

Solution 5

The result of the multiplication:

$$\begin{array}{r} 555\cdots556 \\ 11 \\ \hline 55\ \cdots556 \\ 555\cdots\ 56 \\ \hline 611\ \cdots\ 116 \end{array} \times$$

has 102 digits, of which 100 are equal to 1.

$$\underbrace{55\cdots55}_{100\text{ digits}}6 \times 11 = 6\underbrace{11\cdots11}_{100\text{ digits}}6$$

It is also possible to calculate the result using the fact that $11 = 10+1$. The complexity is the same.

Exercise 6

How many digits of 3 and how many digits of 7 are there in the result of each operation:

Solution 6

(A) $77777 + 5555 = 83332$

There are 3 digits of 3 and 0 digits of 7 in the sum.

(B) $77777777 + 55555 = 77833332$

There are 4 digits of 3 and 2 digits of 7 in the sum.

(C) $\underbrace{77\cdots7}_{\text{20 digits}}+\underbrace{55\cdots5}_{\text{10 digits}}=\underbrace{77\cdots7}_{\text{9 digits}}8\underbrace{33\cdots3}_{\text{9 digits}}2$

There are 9 digits of 3 and 9 digits of 7 in the sum.

(D) $\underbrace{77\cdots7}_{\text{77 digits}}+\underbrace{55\cdots5}_{\text{55 digits}}=\underbrace{77\cdots7}_{\text{21 digits}}8\underbrace{33\cdots3}_{\text{54 digits}}2$

There are 54 digits of 3 and 21 digits of 7 in the sum.

Exercise 7

How many digits of 3 are there in each difference?

Solution 7

(A) $222222-88888=133334$

There are 4 digits of 3 in the difference.

(B) $222222222-888888=221333334$

There are 5 digits of 3 in the difference.

(C) $\underbrace{22\cdots2}_{\text{20 digits}}-\underbrace{88\cdots8}_{\text{19 digits}}=1\underbrace{33\cdots3}_{\text{18 digits}}4$

There are 18 digits of 3 in the difference.

(D) $\underbrace{22\cdots2}_{\text{222 digits}}-\underbrace{88\cdots8}_{\text{88 digits}}=\underbrace{22\cdots2}_{\text{133 digits}}1\underbrace{33\cdots3}_{\text{87 digits}}4$

There are 87 digits of 3 in the difference.

Exercise 8

How many digits of 7 are there in the result of the operation?

$$\underbrace{77\cdots7}_{\text{777 digits}}\underbrace{33\cdots3}_{\text{333 digits}}-\underbrace{55\cdots5}_{\text{555 digits}}=$$

Solution 8

There are $555 + 332 = 885$ digits of 7 in the result:

555 digits 222 digits 333 digits

$$
\begin{array}{r}
77\cdots7\,77\cdots77\,33\cdots333 \\
55\cdots55\,55\cdots555 \\
\hline
77\cdots7\,22\cdots21\,77\cdots778
\end{array}
$$

555 digits 221 digits 332 digits

Exercise 9

If k and m represent numbers of digits, what are the values of k and m?

$$\underbrace{88\cdots8}_{k\text{ digits}}+\underbrace{55\cdots5}_{m\text{ digits}}=\underbrace{88\cdots8}_{12\text{ digits}}9\underbrace{44\cdots4}_{12\text{ digits}}3$$

Solution 9

First, analyze a similar operation with fewer digits:

$$888888 + 555 = 889443$$

Use this information to explore the more difficult case. There are $12 + 1 + 12 + 1 = 26$ digits in the sum. Of these, one digit is a 3 and one digit is a 9. There is one more 5 in the repdigit $\underbrace{55\cdots5}_{m\text{ digits}}$ than there are 4s in the sum. Therefore, there are 13 digits of 5 and $m = 13$.

The sum and the repdigit $\underbrace{88\cdots8}_{k\text{ digits}}$ have the same number of digits: 26. Therefore, $k = 26$.

Exercise 10

If we divide a repdigit formed of digits of 8 by 7 and want the division to be exact, which of the following numbers could be the number of digits of the repdigit?

(A) 7

(B) 8

(C) 12

(D) it is not possible to produce an exact division

Solution 10

Since 8 is not divisible by 7, and:

$$\underbrace{88\cdots8}_{k \text{ digits}} = 8 \times \underbrace{11\cdots1}_{k \text{ digits}}$$

we have to figure out which repunits are divisible by 7. By successive trials we find that a 6-digit repunit is the smallest repunit divisible by 7:

$$111111 = 7 \times 15873 + 0$$

Therefore, we also have:

$$888888 = 7 \times 126984 + 0$$

Thus, a 6-digit repdigit with digits of 8 is exactly divisible by 7. This is true for all numbers of digits that are multiples of 6. Choice (C) is correct.

Solutions to Practice Four

Exercise 1

Find all the pairs of integers (m, n) that satisfy the equality:

$$77 = (m + 4)(n - 4)$$

Solution 1

Factor 77 into primes. Since the prime factorization of any number is unique, the factors must be the same on both sides of the equation.

$$7 \times 11 = (m + 4)(n - 4)$$

Two choices are possible:

$$\begin{aligned} m + 4 &= 7 \\ n - 4 &= 11 \end{aligned}$$

and

$$\begin{aligned} m + 4 &= 11 \\ n - 4 &= 7 \end{aligned}$$

There are two solutions for the pair (m, n): $(3, 15)$ and $(7, 11)$.

Note: this is a quadratic (m and n are multiplied) Diophantine equation with 2 unknowns.

Exercise 2

Find all the integer(s) k that satisfy the equality:

$$(k-5)\times(k+3)=33$$

Solution 2

Factor 33 into primes:

$$(k-5)\times(k+3)=3\times 11$$

$k=8$ satisfies both equations below:

$$\begin{aligned} k-5 &= 3 \\ k+3 &= 11 \end{aligned}$$

k is equal to 8. Because $k+3>k-5$ for any k, there is no value of k that can satisfy the following equations:

$$\begin{aligned} k-5 &= 11 \\ k+3 &= 3 \end{aligned}$$

Therefore, there is only one solution $k=8$.

Note: this is a quadratic Diophantine equation with 1 unknown.

Exercise 3

Find the value of x:

$$((((((x+4)-5)+6)-7)+8)-9)+10=3$$

Solution 3

The parentheses are superfluous since they separate only terms of additions (addition is commutative). Simply discard them:

$$x+4-5+6-7+8-9+10=3$$

and notice that each pair of numbers from right to left has a sum of 1:

$$x + 4 + 1 + 1 + 1 = 3$$

x must equal -4.

Note: this is a linear equation with 1 unknown.

Exercise 4

Find the value of the unknown quantity x:

$$x + 4 - (5 + 6 - (7 + 8 - (9 + 10))) - 4$$

Solution 4

$$\begin{aligned}
x + 4 - (5 + 6 - (7 + 8 - (9 + 10))) &= 4 \\
x - (5 + 6 - (7 + 8 - (9 + 10))) &= 0 \\
x &= (5 + 6 - (7 + 8 - (9 + 10))) \\
x &= 11 - (15 - (19)) \\
x &= 11 - (-4) \\
x &= 11 + 4 \\
x &= 15
\end{aligned}$$

Note: this is a linear equation with 1 unknown.

Exercise 5

What is the digit replaced by the $\clubsuit$?

$$\underbrace{\heartsuit\heartsuit\cdots\heartsuit}\times 7 = 6\underbrace{\clubsuit\clubsuit\cdots\clubsuit}16$$

Solution 5

To obtain a last digit of 6 in the product, 7 must be multiplied by 8. Therefore, the $\heartsuit$ is an 8. We notice that, to obtain a digit of $\clubsuit$ in the product, the number of digits in the repunit $\heartsuit\heartsuit\cdots\heartsuit$ must be at least equal to three. Performing any of the multiplications:

$$\begin{aligned} 888\times 7 &= 6216 \\ 8888\times 7 &= 62216 \\ \text{etc.} \end{aligned}$$

shows that $\clubsuit = 2$.

Note: this is a Diophantine equation where the unknowns are digits, also known as a *cryptarithm*.

Exercise 6

Ali and Baba discovered an underground palace and set out to map all the treasure inside it. After carefully mapping the interior, they found that the palace had a vault filled with treasure chests. In each chest, they found either 7 gold bars or 11 silver bars. There were 90 bars in total. How many of them were gold bars?

Solution 6

The number of silver bars must be a multiple of 11. The number of gold bars must be a multiple of 7. Since 90 is *even*, the number of gold bars and the number of silver bars must have the same parity (they are either both even or both odd).

If both numbers are even, then we can divide the total number of bars by 2 and get that 45 must be the sum of a multiple of 11 and a multiple of 7. It is easy to check that 11, 22, and 33 do not satisfy (44 is clearly too large).

If both numbers are odd, then we can use the odd multiples of 11:

$$\begin{aligned}
90 - 11 &= 79 \quad \text{not a multiple of 7} \\
90 - 33 &= 57 \quad \text{not a multiple of 7} \\
90 - 55 &= 35 \quad \text{a multiple of 7} \\
90 - 77 &= 13 \quad \text{not a multiple of 7}
\end{aligned}$$

There are 55 silver bars and 35 gold bars.

Note: this problem is modeled using a linear Diophantine equation with 2 unknowns.

Exercise 7

When Lila divides a number by 11, she gets the same result as when she subtracts 80 from it. What is the number?

Solution 7

Solution 1: The number is equal to 11 times the number minus 880. Therefore, 10 times the number is 880. The number is 88.

Solution 2: Set up an equation:

$$\begin{aligned}
\frac{N}{11} &= N - 80 \\
N &= 11 \times N - 880 \\
880 &= 10 \times N
\end{aligned}$$

$$N = 88$$

Note: this problem is modeled using a linear equation with 1 unknown.

Exercise 8

Dina helped organize a raffle at the farmer's market. She packaged apples in bags of 5 and quinces in bags of 4. In total, she packaged 111 apples and quinces. There were fewer quinces than apples but fewer apples than twice the quinces. How many bags of each type of fruit did she package?

Solution 8

The number of quinces must be a multiple of 4 and the number of apples must be a multiple of 5. Since the number of apples and quinces combined is *odd* and the number of quinces is *even*, then the number of apples must be *odd*.

Since the number of apples is an odd multiple of 5, its last digit is 5. It follows that the last digit of the number of quinces must be 6. Multiples of 4 that end in 6 are: 16, 36, 56, etc. As the number of quinces increases, it surpasses the number of apples. This allows us to stop looking for more solutions.

$$\begin{aligned}
4 \times 4 &= 16, \quad 111 - 16 = 95, \quad 95 = 5 \times 19 \\
4 \times 9 &= 36, \quad 111 - 36 = 75, \quad 75 = 5 \times 15 \\
4 \times 14 &= 56, \quad 111 - 56 = 55, \quad 55 = 5 \times 11
\end{aligned}$$

We can stop now since the number of quinces (56) is larger than the number of apples (55).

The solutions are $(4, 19)$ and $(9, 15)$. Of these, only $(9, 15)$ satisfies the inequality relations in the problem (there are fewer apples than twice the quinces).

Note: this problem is modeled using a linear Diophantine equation with 2 unknowns.

Exercise 9

Which triangular number is also a 3-digit repdigit?

Solution 9

Triangular numbers are calculated by giving positive integer values to N in the formula:

$$\frac{N \times (N+1)}{2}$$

A 3-digit repdigit is a positive integer of the form mmm, where m is a digit between 1 and 9.

The problem asks us to solve:

$$\frac{N \times (N+1)}{2} = m \times 111$$

This is a Diophantine equation which can be solved by factoring and by considering the limited range of values that m can have.

$$\frac{N \times (N+1)}{2} = m \times 3 \times 37$$
$$N \times (N+1) = 2 \times 3 \times 37 \times m$$

The product on the right hand side must be the product of two consecutive numbers. Since 37 is prime, we are looking for a possible value of m that would make the product of the remaining factors equal either 36 or 38. Only 36 can be achieved if $m = 6$: $2 \times 3 \times 6$. No other solutions are possible.

The triangular number is 666.

Exercise 10

Find the value of x.

Solution 10

$$\frac{\cancel{5}}{x} \times \frac{\cancel{4}}{\cancel{5}} \times \frac{\cancel{3}}{\cancel{4}} \times \frac{2}{\cancel{3}} = \frac{1}{4}$$
$$\frac{2}{x} = \frac{1}{4}$$

x must be 8.

Exercise 11

Find the value of x and write it in the form $n+\frac{a}{b}$ where n is an integer and $\frac{a}{b}$ is a proper irreducible fraction.

Solution 11

$$\begin{aligned}
\frac{1}{x} &= \frac{3}{6} \times \frac{4}{7} \times \frac{5}{8} \times \frac{6}{9} \\
&= \frac{\not{3}}{\not{6}} \times \frac{\not{4}}{7} \times \frac{5}{2 \times \not{4}} \times \frac{\not{6}}{3 \times \not{3}} \\
&= \frac{5}{7 \times 2 \times 3} \\
&= \frac{5}{42}
\end{aligned}$$

$$x = \frac{42}{5} = 8 + \frac{2}{5}$$

Note: Never simplify like this!

$$\frac{\not{3}+2}{\not{3}+5}$$

A simple calculation can prove that it is incorrect to simplify in this manner.

Solution 12

(A) Instead of adding 17 to a number, Dina subtracted it. Which number should she add to the result to get the intended correct answer? 34

(B) Instead of subtracting 101 from a number, Lila added it. Which number should she subtract from the result to get the intended correct answer? 202

(C) Instead of multiplying a number by 12, Amira divided it by 4. What operation is needed for her to obtain the intended correct answer? **Amira should multiply the result by 48.**

(D) Instead of dividing a number by 4, Dina multiplied it by 16. What operation is needed for her to obtain the intended correct answer? **Dina should divide the result by 64.**

Exercise 13

Amira wanted to make a larger cube out of 12 small cubes. When she finished, she realized she had not used all the small cubes. Which of the following could be the number of small cubes left over? (check all that apply)

(A) 0
(B) 2
(C) 3
(D) 4

Solution 13

The number of small cubes used must be a *cube number*. A cube number is the result of multiplying a number by itself two times. The cube numbers larger than 0 and smaller than 12 are:

$$1 = 1 \times 1 \times 1$$
$$8 = 2 \times 2 \times 2$$

Since the cube she built is larger than a small cube, only the last option is possible. The number of leftover cubes must be:

$$12 - 8 = 4$$

The correct answer is (D).

Exercise 14

Find the positive integers A, B, and C, if:

$$\begin{aligned} A+B+C &= 50 \\ B+B+C &= 38 \\ C+C+C &= 36 \end{aligned}$$

Solution 14

Start with the third equation which uses only C. Since $3 \times C = 36$, $C = 12$.

Continue with the middle equation where only B and C are used. You now know the value of C: $2 \times B = 38 - 12$, $2 \times B = 26$. Therefore, $B = 13$.

Finally, solve the first equation:

$$A + 13 + 12 = 50$$

$A = 25$.

Exercise 15

Find the positive integers A, B, and C, if:

$$\begin{aligned} A \times B \times C &= 210 \\ B \times B \times C &= 252 \\ C \times C \times C &= 343 \end{aligned}$$

Solution 15

Start with the third equation which uses only C. Factor the number 343 into primes and find that $7 \times 7 \times 7 = 343$. Therefore, $C = 7$.

Continue with the middle equation where only B and C are used, but now you know the value of C: $C \times C \times 7 = 252$. Divide 252 by 7: $252 \div 7 = 36$. Therefore, $B = 6$.

Finally, use the first equation since you now know the values of both B and C:

$$A \times 6 \times 7 = 210$$

$A = 5$.

Exercise 16

Find the numbers A, B, and C, if:

$$\begin{aligned} A + B &= 25 \\ B + C &= 38 \\ A + C &= 47 \end{aligned}$$

Solution 16

This is a typical system of equations in which the solution hinges on the observation that, by adding all three equations together, we get:

$$2 \times A + 2 \times B + 2 \times C = 25 + 38 + 47 = 110$$

It follows that:

$$A + B + C = 55$$

Since $A + B = 35$, $C = 55 - 35 = 20$.

Since $B + C = 38$, $A = 55 - 38 = 17$.

Since $A + C = 47$, $B = 55 - 47 = 8$.

Solutions to Practice Five

Exercise 1

A 40 cm stick is what fraction of 1 meter long?

Solution 1

Convert 1 meter to centimeters. Since 1 m = 100 cm, the fraction is:

$$\frac{40}{100} = \frac{4}{10} = \frac{2}{5}$$

Exercise 2

Fill in the missing values in the table:

Solution 2

Length	30 cm	40 cm	75 cm	10 mm	300 mm
Fraction of 1 meter	$\frac{3}{10}$	$\frac{2}{5}$	$\frac{3}{4}$	$\frac{1}{100}$	$\frac{3}{10}$

Exercise 3

Dina and Lila were throwing a summer party. They had ten 750 milliliter (mL) bottles of lemonade and twelve 1 liter (L) bottles of limeade. They filled glasses with 150 milliliters of either lemonade or limeade until all the bottles were empty. How many more glasses of limeade than of lemonade were there when they finished?

Solution 3

The number of glasses of lemonade was:

$$10 \times 750 \div 150 = 50$$

The number of glasses of limeade was:

$$12 \times 1000 \div 150 = 80$$

There were 30 more glasses of limeade than glasses of lemonade. Note how it was necessary to convert the volume of 1L to 1000 mL in order to have the same units throughout the operation.

Exercise 4

5 hours and 20 minutes are what fraction of one day?

Solution 4

Let us convert all the time intervals to minutes. 5 hours and 40 minutes represent:

$$5 \times 60 + 20 = 320 \text{ minutes}$$

One day represents:

$$24 \times 60 = 1440 \text{ minutes}$$

The given interval represents $\frac{320}{1440}$ of a day. This fraction can be further simplified to:

$$\frac{320}{1440} = \frac{\cancel{2} \times \cancel{2} \times \cancel{2} \times \cancel{2} \times \cancel{2} \times 2 \times \cancel{5}}{\cancel{2} \times \cancel{2} \times \cancel{2} \times \cancel{2} \times \cancel{2} \times 3 \times 3 \times \cancel{5}} = \frac{2}{9}$$

Exercise 5

A side of a square is 40 cm long. Dina divides the square into 7 squares of different sizes. What is the side length of the smallest square?

Solution 5

Dina divided the large square like this:

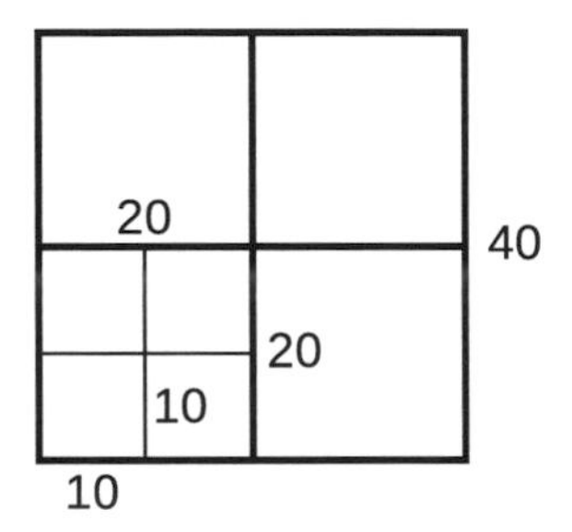

The smallest square has a side length of 10 cm.

Exercise 6

What fraction of the perimeter of a square with a side length of 20 units are the perimeters of the following figures:

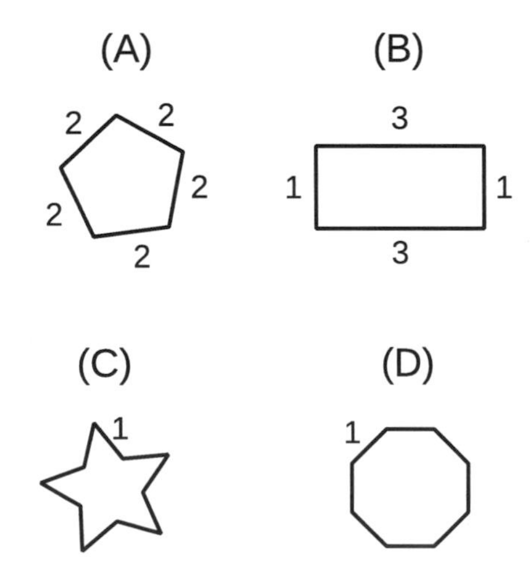

Solution 6

The perimeter of a square with a side length of 20 units is equal to 80 units.

The perimeter of the figures are:

(A) $5 \times 2 = 10$ units, one eighth of the perimeter of the square

(B) $3 \times 2 + 1 \times 2 = 8$ units, one tenth of the perimeter of the square

(C) $1 \times 10 = 10$ units, one eighth of the perimeter of the square

(D) $1 \times 8 = 8$ units, one tenth of the perimeter of the square

Exercise 7

Stephan, the tennis coach, must drive to a tournament. One of his students is also going, but in a separate vehicle. Stephan drives at a speed of 108 kilometers per hour, while the student drives at a speed of 30 meters per second. Who is driving faster?

(A) Stephan

(B) The student

(C) Both drive at the same speed.

Solution 7

To compare the quantities we must convert them to a single unit. For illustration, let us do both conversions.

Convert to meters per second:

$$\begin{aligned} 108\ \frac{\text{km}}{\text{hr}} &= 108\ \frac{1000\ \text{m}}{3600\ \text{s}} \\ &= 108\ \frac{10\ \text{m}}{36\ \text{s}} \\ &= \frac{\cancel{36} \times 3 \times 10}{\cancel{36}} \frac{\text{m}}{\text{s}} \\ &= 30\ \frac{\text{m}}{\text{s}} \end{aligned}$$

Convert to kilometers per hour:

$$\begin{aligned} 30 \frac{\text{m}}{\text{s}} &= 30 \frac{\text{km}}{1000} \frac{3600}{1\ \text{hr}} \\ &= \frac{30 \times 36}{10} \frac{\text{km}}{\text{hr}} \\ &= 108 \frac{\text{km}}{\text{hr}} \end{aligned}$$

The correct answer is (C).

Exercise 8

Amira decided to help her mother bake muffins. She measured 300 mL of milk in a measuring cup. Then, she put 6 squares of chocolate in the cup. After she added chocolate, the level in the measuring cup rose to 380 mL. What was the volume of the chocolate?

Solution 8

The volume of the chocolate was $380 - 300 = 80$ mL. This method of measuring the volume of irregularly shaped objects is called *measuring a volume by displacement.* To use this method, the object to be measured must be completely immersed in the liquid.

Exercise 9

For a school project, Dina and Lila had to measure a rectangular kitchen table. Dina measured the perimeter using a pencil as a unit and got a length of 34. Lila measured the perimeted using a bamboo skewer as a unit and got a length of 29. Which was longer, the pencil or the skewer?

Solution 9

The skewer was longer. Since the same length was measured using both methods, the shorter unit resulted in a larger number of units for the total length.

Exercise 10

A square has a perimeter of 29 meters. Amira wanted to find a unit so that she would obtain a whole number of units when she measured the side of the square. Which of the following is a possible length of the unit?

(A) 8 cm
(B) 33 cm
(C) 58 cm
(D) 145 cm

Solution 10

29 meters = 2900 cm. Each side of the square has a length of $2900 \div 4 =$ 725 cm. The prime factorization of 725 is:

$$725 = 5 \times 5 \times 29$$

The side of the square will be expressed as a whole number for any unit that is a divisor of 725 cm. The divisors of 725 are: 1, 5, 25, 29, 145, and 725. The only answer choice that is among the divisors is 145. The correct answer is (D).

Exercise 11

Ali wanted to get his donkey to return home with him from the Market. The donkey had a will of its own and did not want to go home without quenching his thirst, so they had to go to the Fountain first. The Fountain, however, is not on the street that leads from the Market to Ali's home, as the map in the figure shows:

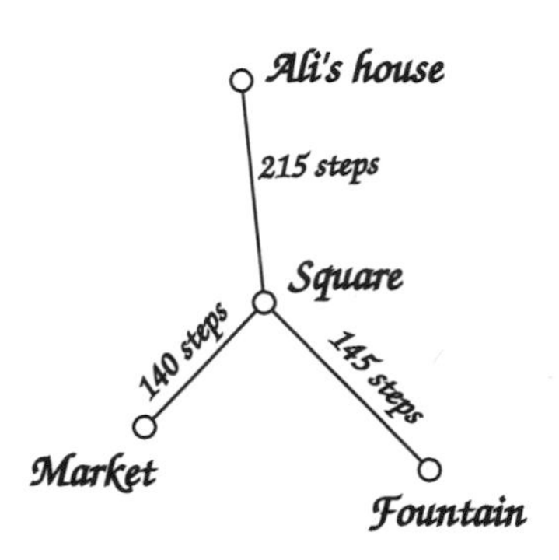

The distances shown on the map are measured in steps. How many steps did the donkey travel on its way from the Market to Ali's house?

Solution 11

The donkey traveled on the route: Market-Square-Fountain-Square-Ali's house. It covered a total distance of: $140 + 145 + 145 + 215 = 645$ steps.

Exercise 12

Dina's mother planted a square patch of raspberry bushes in the garden. Over the years, the raspberry plants grew outside the square. One day, Dina's mother noticed that, while the patch was still square, its perimeter had tripled. She asked Dina to find out how many times larger than before the area of the patch was. What did Dina find?

Solution 12

Dina thought, "If the perimeter tripled and the patch is still square, then each side of the small square must have tripled:"

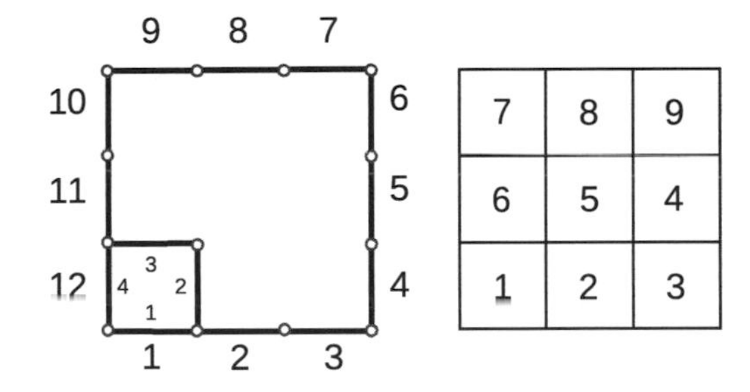

Dina noticed that, if the perimeter is 3 times larger, then the area is 9 times larger.

Solutions to Miscellaneous Practice

Solution 1

$$\frac{1}{2}+\frac{1}{3}+\frac{1}{4}+\frac{1}{5}=\frac{77}{60}$$

Solution 2

$$\begin{aligned}
\frac{11}{4} &= \frac{5}{4}+\frac{6}{4}\\
\frac{23}{7} &= \frac{11}{7}+\frac{12}{7}\\
\frac{211}{42} &= \frac{105}{42}+\frac{106}{42}\\
\frac{901}{17} &= \frac{450}{17}+\frac{451}{17}
\end{aligned}$$

Exercise 3

Find the solution of the cryptarithm:

```
 AARD
  ARD
   RD
    D
-----  +
KMAMP
```

Solution 3

To obtain a fifth digit in the result, A must be equal to 9 and there must be a carryover from the column $A + A$. Since that column must reproduce A, there must be a carryover of 1 from the column $R+R+R$. M must be equal to 0. Since the $R + R + R$ column produces a result that ends in zero, R must be equal to 3 and $D + D + D + D$ must produce a carryover of 1. Only the digits 3 and 4 produce carryovers of 1 when multiplied by 4. Since 3 has been used already, D must be equal to 4:

$$\begin{array}{r} 9934 \\ 934 \\ 34 \\ 4 \\ \hline 10906 \end{array} +$$

Exercise 4

Alfonso, the grocer, received a shipment of 95 unsorted avocadoes. He found that he was able to package the smaller avocadoes in bags of 9 each and the larger avocadoes in bags of 4 each, without any avocadoes left over. How many small avocadoes were in the shipment?

Solution 4

Since the number of avocadoes of each kind is a positive integer, this problem can be modeled using a Diophantine equation.

If S is the number of bags of small avocadoes and L is the number of bags of large avocadoes, we know that:

$$4 \times L + 9 \times S = 95$$

Since $4 \times L$ is even, $9 \times S$ must be odd in order to get a sum of 95.

There are 5 odd multiples of 9 smaller than 95: 9, 27, 45, 63, and 81.

Which ones will give us multiples of 4 when subtracted from 95?

$$
\begin{aligned}
95 - 81 &= 14 \\
95 - 63 &= \mathbf{32} \\
95 - 45 &= 50 \\
95 - 27 &= \mathbf{68} \\
95 - 9 &= 86
\end{aligned}
$$

There were either 27 or 63 small avocadoes.

The number of bags used in each case was:

Case 1: 27 small avocadoes fit into 3 bags. The remaining 68 large avocadoes fit into 17 bags. The total number of bags is 20.

Case 2: 63 small avocadoes fit into 7 bags. The remaining 32 large avocadoes fit into 8 bags. The total number of bags is 15.

To maximize the total number of bags, we select the solution with 27 small avocadoes.

Exercise 5

If two circles have the same weight as three triangles and two triangles have the same weight as one square, which object should be removed from the balance to establish equilibrium?

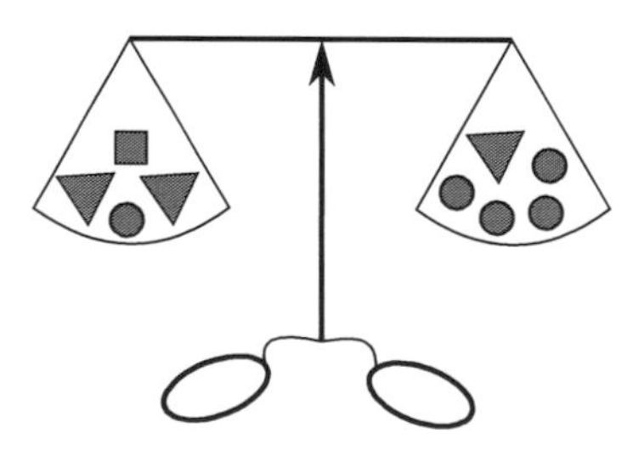

Solution 5

Remove 1 triangle and 1 circle from each pan.

There are 3 circles left in the right pan. Exchange 2 of the circles with 3 triangles and 2 of the triangles with 1 square. Now we have 1 square, 1 triangle, and 1 circle in the right pan and 1 square and 1 triangle in the left pan.

To balance the scale, one circle must be removed from the right pan. The correct answer is (A).

Solution 6

Compute efficiently:

1. 999×1001

$$\begin{aligned} &= 999 \times 1000 + 999 \\ &= 999000 + 999 \\ &= 999999 \end{aligned}$$

2. 99×1281

$$\begin{aligned} &= 100 \times 1281 - 1281 \\ &= 128100 - 1281 \\ &= 126819 \end{aligned}$$

3. 13×125

$$\begin{aligned} &= 13 \times 125 \times 8 \div 8 \\ &= 13 \times 1000 \div 8 \\ &= 13000 \div 8 \\ &= 1625 \end{aligned}$$

In this operation, we consider division by a 1-digit number to be less complex than multiplication of multi-digit numbers. This is especially true when we compute or estimate in our heads.

4. $ab \times 101$

$$\begin{aligned} &= ab \times 100 + ab \\ &= ab00 + ab \\ &= abab \end{aligned}$$

This is a pattern of digits that is common in problems. Regardless of which digits a and b are, a number of the form $abab$ is always divisible by 101.

5. 64×75

$$\begin{aligned} &= 16 \times 4 \times 25 \times 3 \\ &= 16 \times 100 \times 3 \\ &= 16 \times 3 \times 100 \\ &= 4800 \end{aligned}$$

Notice that it is sometimes more efficient to factor the numbers before multiplying them. Then, we can multiply the smaller factors by choosing the order of the multiplications in a clever way.

6. 9999×9999

$$\begin{aligned} &= 9999 \times 10000 - 9999 \\ &= 99990000 - 9999 \\ &= 99980001 \end{aligned}$$

Exercise 7

How many shaded squares are there?

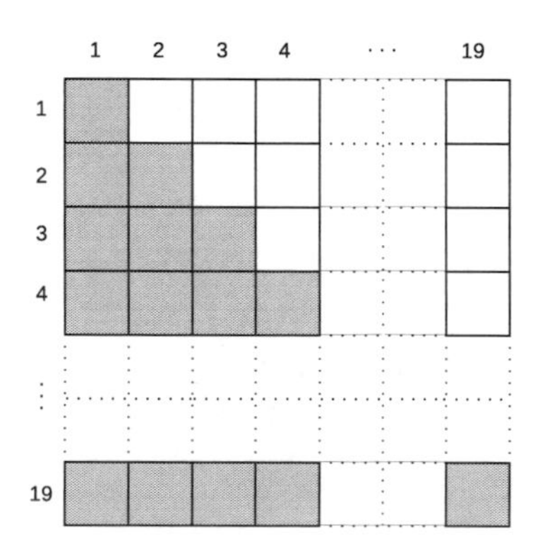

Solution 7

There are 190 shaded squares.

Strategy 1: Subtract the number of squares on the diagonal from the total number of squares, divide the result by 2 and add back the diagonal:

$$19 \times 19 - 19 = 19 \times 18$$

$$19 \times 18 \div 2 = 19 \times 9$$

$$19 \times 9 + 19 = 19 \times 10 = 190$$

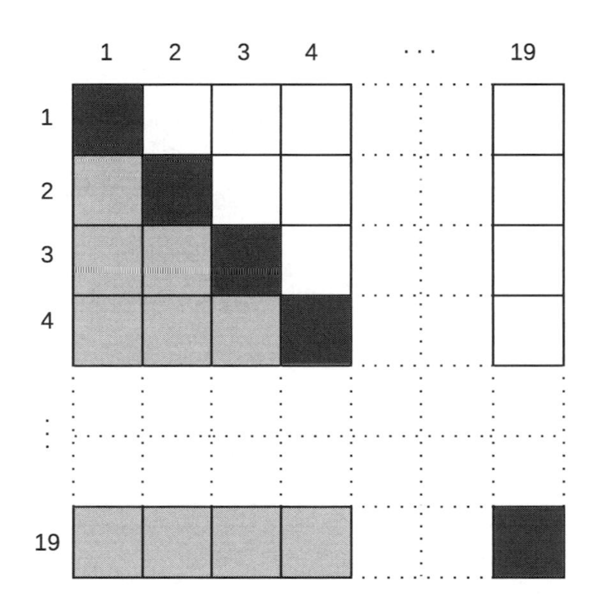

Strategy 2: The number of shaded squares is a triangular number:

$$1 + 2 + 3 + \cdots + 19 = \frac{19 \times 20}{2} = 19 \times 10 = 190$$

Exercise 8

Find x:

$$1 - (2 - (3 - (4 - (5 - (6 - (7 - (8 - x))))))) = 0$$

Solution 8

Strategy 1: A minus sign placed in front of a parenthesis flips all the operations inside the parentheses: minus becomes plus and plus becomes minus. **Attention:** numbers without a sign in front of them are positive.

Count the number of flips applied to each term throughout the operations. Two flips cancel each other. Numbers with an even number of flips keep their sign. Numbers with an odd number of flips change their sign.

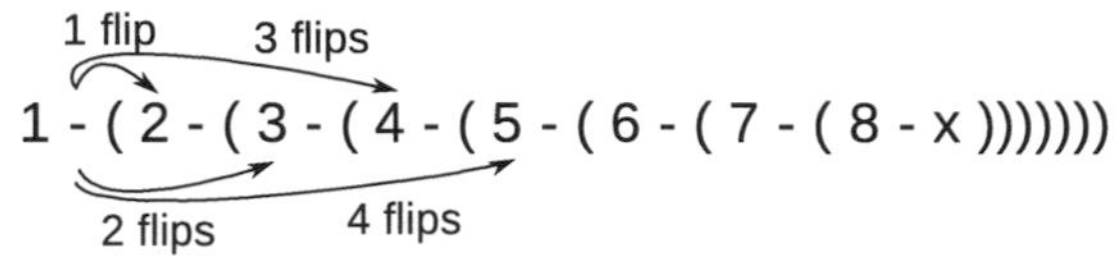

1 flip 3 flips
1 - (2 - (3 - (4 - (5 - (6 - (7 - (8 - x)))))))
2 flips 4 flips

Now, let's put this together:

$$1 - 2 + 3 - 4 + 5 - 6 + 7 - 8 + x = 0$$

$$-1 - 1 - 1 - 1 + x = 0$$

Therefore, $x = 4$.

Strategy 2: Notice that if we set x equal to 4 it all unravels "magically" to zero:

$$\begin{aligned}
8-4 &= 4\\
7-4 &= 3\\
6-3 &= 3\\
5-3 &= 2\\
4-2 &= 2\\
3-2 &= 1\\
2-1 &= 1\\
1-1 &= 0
\end{aligned}$$

Exercise 9

Place parentheses to make the expression an identity:

$$4 \times 4 - 4 \div 4 + 4 \times 4 = 4 \times 4 \times 4$$

Solution 9

$$\begin{aligned}
(4 \times (4 - 4 \div 4) + 4) \times 4 &= 4 \times 4 \times 4\\
(4 \times (4 - 1) + 4) \times 4 &= 4 \times 4 \times 4\\
(4 \times 3 + 4) \times 4 &= 4 \times 4 \times 4\\
(12 + 4) \times 4 &= 4 \times 4 \times 4\\
16 \times 4 &= 4 \times 4 \times 4
\end{aligned}$$

Exercise 10

Place parentheses so that the result of the operations is an integer as large as possible:

$$5 \times 8 + 12 \div 3 + 5$$

Solution 10

$$\begin{aligned} & 5 \times (8 + 12 \div 3 + 5) \\ = & 5 \times (8 + 4 + 5) \\ = & 5 \times 17 \\ = & 85 \end{aligned}$$

Exercise 11

In the figure, each brick has a number that is the sum of the numbers on the two bricks below it. Which number is on the brick marked with an "X?"

Solution 11

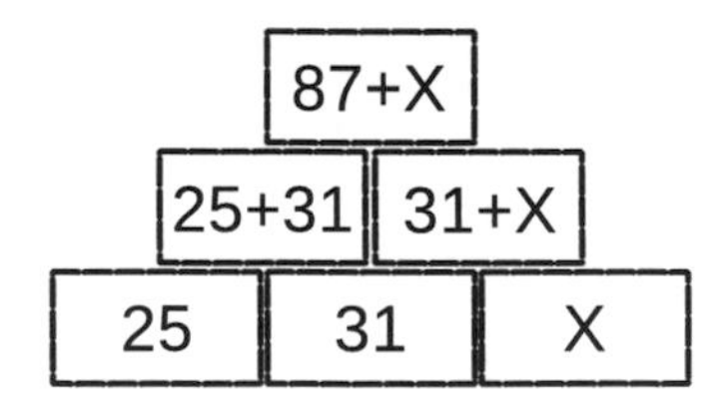

$x = 13$

Exercise 12

In the figure, each brick has a number that is the product of the numbers on the two bricks below it. None of the numbers is equal to 1. Which number is on the brick marked with an "X?"

Solution 12

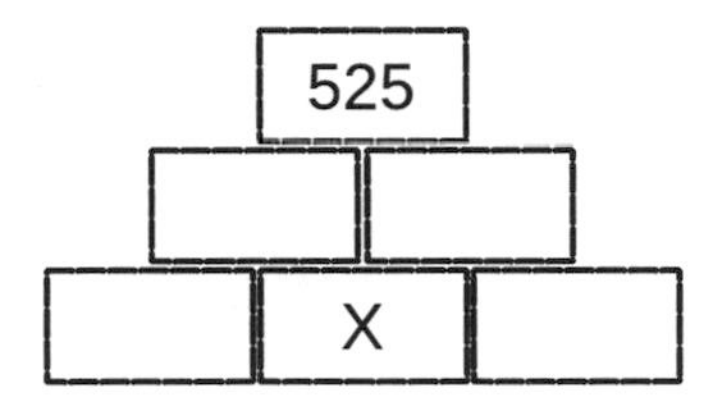

Factor 525 into primes:

525	3
175	5
35	5
7	7
1	

525 = 3 x 5 x 5 x 7

and match the result with the scheme of operations:

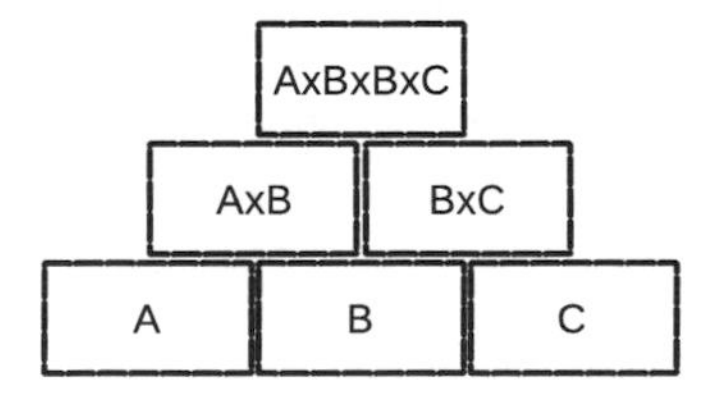

The number 5 is written on the brick marked with an "X."

Exercise 13

Lila and Dina played a game of soccer while on opposing teams. The difference between the scores of the two teams was 4 times less than the total number of points scored. The sum of the total number of points scored and the difference between the scores was 40. If Lila was on the losing team, what score did her team have?

Solution 13

Denote the scores a and b and assume a is the larger score. The sum of the scores is $a + b$. The difference of the scores is $a - b$.

$$\begin{aligned} a + b + a - b &= 40 \\ a + a &= 40 \\ a &= 20 \end{aligned}$$

The difference between the scores was 4 times less than the total number of points scored:

$$\begin{aligned} 4 \times (a - b) &= (a + b) \\ 4 \times (20 - b) &= (20 + b) \\ 80 - 4 \times b &= 20 + b \\ 60 &= 5 \times b \\ b &= 12 \end{aligned}$$

Lila's team scored 12 points.

Exercise 14

How many 1 digits are there in the result of the operation:

$$\underbrace{10101 \cdots 10}_{100 \text{ digits}} - \underbrace{9090 \cdots 909}_{99 \text{ digits}}$$

Solution 14

Work on similar, but less complex operations first:

$$\begin{aligned} 1010 - 909 &= 101 \\ 101010 - 90909 &= 10101 \\ \cdots &= \cdots \\ \underbrace{10101\cdots10}_{100 \text{ digits}} - \underbrace{9090\cdots909}_{99 \text{ digits}} &= \underbrace{1010\cdots101}_{99 \text{ digits}} \end{aligned}$$

There are 45 digits of 1 and 44 digits of 0 in the result.

Exercise 15

How many digits of 1 are there in the result of the operation:

$$\underbrace{10101\cdots10}_{100 \text{ digits}} - \underbrace{9090\cdots909}_{49 \text{ digits}}$$

Solution 15

```
                49th place
                    ↓
1 0 1 0 ... 1 0 1 0 1 0 1 ...0 1 0 1 0
                    9 0 9 0... 9 0 9 0 9
------------------------------------------ -
1 0 1 0 ...1 0 0 1 0 1 0 ...1 0 1 0 1
                  ↑
              50th place
```

Among the last 50 places, 25 are 1s and 25 are 0s. Among the first 50 places, 25 are 1s and 25 are 0s. The answer has a total of 50 digits of 1 and 50 digits of 0.

Exercise 16

Find the consecutive positive integers k, l, and m:

$$(k-1)(l-1)(m-1) = 504$$

Solution 16

Factor the right hand side into primes:

$$504 = 2 \times 2 \times 2 \times 3 \times 3 \times 7$$

Then, find a convenient grouping of the factors so you obtain three consecutive factors:

$$(2 \times 2 \times 2) \times (3 \times 3) \times 7 = 7 \times 8 \times 9$$

k, l, and m are 8, 9, and 10, in no particular order. (We are told they are consecutive but we are not told how they are ordered.)

Exercise 17

Find a in each of the following:

1. $(a - 2.55 \div 3) \times 8 = \frac{6}{5}$
2. $450 - 171 \div a = 441$
3. $(41 \times a - 497) \div 149 = 14$
4. $4 \div (5 - 4 \div (3 + 3 \div a)) = 1$
5. $5 \times (a \times 1 \div a + a \times 0 - a \div 1 - 6 \times a \div 2) \div 11 = 95 \div 209$

Solution 17

In all the equations, we use strategies that *increase the number of steps* while *decreasing the complexity of the operations.* Instead of multiplying large numbers, factor them out and simplify. Instead of manipulating decimals, use fractions.

1.

$$
\begin{aligned}
(a - 2.55 \div 3) \times 8 &= \frac{6}{5} \\
(a - 2.55 \div 3) \times 40 &= 6 \\
(a - 2.55 \div 3) \times 40 &= 6 \\
40 \times a - 40 \times 2.55 \div 3 &= 6 \\
40 \times a - 4 \times 25.5 \div 3 &= 6 \\
40 \times a - 102 \div 3 &= 6 \\
40 \times a - 34 &= 6 \\
40 \times a &= 6 + 34 \\
40 \times a &= 40 \\
a &= 1
\end{aligned}
$$

2.

$$
\begin{aligned}
450 - 171 \div a &= 441 \\
450 &= 441 + 171 \div a \\
450 - 441 &= 171 \div a \\
9 &= 171 \div a \\
9 \times a &= 171 \\
a &= 171 \div 9 \\
a &= 19
\end{aligned}
$$

3.

$$\begin{aligned}
(41 \times a - 497) \div 149 &= 14 \\
41 \times a - 497 &= 14 \times 149 \\
41 \times a &= 14 \times 149 + 497 \\
41 \times a &= 14 \times 149 + 7 \times 71 \\
41 \times a &= 14 \times (150 - 1) + 7 \times 71 \\
41 \times a &= 7 \times (300 - 2) + 7 \times 71 \\
41 \times a &= 7 \times (300 - 2 + 71) \\
41 \times a &= 7 \times (371 - 2) \\
41 \times a &= 7 \times 369 \\
41 \times a &= 7 \times 41 \times 9 \\
a &= 63
\end{aligned}$$

4.

$$\begin{aligned}
4 \div (5 - 4 \div (3 + 3 \div a)) &= 1 \\
5 - 4 \div (3 + 3 \div a) &= 4 \\
5 - 4 \div (3 + 3 \div a) - 4 &= 0 \\
1 - 4 \div (3 + 3 \div a) &= 0 \\
1 &= 4 \div (3 + 3 \div a) \\
3 + 3 \div a &= 4 \\
3 - 4 + 3 \div a &= 0 \\
3 \div a &= 1 \\
a &= 3
\end{aligned}$$

5. Note: 209 is divisible by 11 since $9 - 0 + 2 = 11$.

$$\begin{aligned}
5 \times (a \times 2 \div a + a \times 0 - a \div 1 - 6 \times a \div 2) \div 11 &= 95 \div 209 \\
5 \times (2 + 0 - a - 3 \times a) &= 5 \times 19 \times 11 \div 209 \\
5 \times (2 - 4 \times a) &= 5 \times 19 \times 11 \div (11 \times 19) \\
2 - 4 \times a &= 19 \times 11 \div 11 \div 19 \\
2 - 4 \times a &= 19 \div 19 \\
2 - 4 \times a &= 1 \\
2 - 1 &= 4 \times a \\
1 &= 4 \times a \\
a &= \frac{1}{4}
\end{aligned}$$

Exercise 18

Amira has learned how to add, subtract, multiply, and divide multi-digit numbers. Because she is still little, it takes her some time to compute. She can perform an addition in 1 minute and a multiplication in 3 minutes. How quickly can Amira find the result of the operations:

$$3 \times 30 + 5 \times 30$$

Solution 18

Amira can find the result in only 4 minutes:

$$(3 + 5) \times 30$$

She adds $3 + 5$ and multiplies the result by 30.

Exercise 19

Lila and Amira go to the same school. The product of their ages is 143. How many years older than Amira is Lila?

Solution 19

The ages are positive integers and must be factors of 143. Since $143 = 11 \times 13$, Amira must be 11 and Lila must be 13. Lila is 2 years older than Amira.

Exercise 20

Lila, Dina, and Amira use a shared piggy bank to save money for toys and jewelry. Before the winter holidays, they decided to use the money they had saved. Dina spent one quarter of the money on a bracelet, Lila spent one third of the remaining money on little pets, and Amira spent half of the remaining money on talking beads. Which of the following is true:

(A) the talking beads were the most expensive

(B) the bracelet was the most expensive

(C) the little pets were the most expensive

(D) they all cost the same

Solution 20

After Dina used one quarter of the money, three quarters were left. Lila used a third of three quarters, which is one quarter of the initial amount. After Dina and Lila each used one quarter, half the initial amount was left in the bank. Of this half, Amira used one half, which was one quarter of the initial amount. Therefore, each of the girls spent the same amount of money. The correct answer is (D).

Competitive Mathematics Series for Gifted Students

Practice Counting (ages 7 to 9)
Practice Logic and Observation (ages 7 to 9)
Practice Arithmetic (ages 7 to 9)
Practice Operations (ages 7 to 9)

Practice Word Problems (ages 9 to 11)
Practice Combinatorics (ages 9 to 11)
Practice Arithmetic(ages 9 to 11)
Practice Operations (ages 9 to 11)

Practice Word Problems (ages 11 to 13)
Practice Combinatorics (ages 11 to 13)
Practice Arithmetic and Number Theory (ages 11 to 13)
Practice Algebra and Operations (ages 11 to 13)
Practice Geometry (ages 11 to 13)

Practice Word Problems (ages 12 to 15)
Practice Algebra and Operations (ages 12 to 15)
Practice Geometry (ages 12 to 15)
Practice Number Theory (ages 12 to 15)
Practice Combinatorics and Probability (ages 12 to 15)

This is a series of practice books. With the exception of a few reminders, there are no theoretical explanations. For lessons, please see the resources indicated below:

Find a set of free lessons in competitive mathematics at www.mathinee.com. Addressing grades 5 through 11, the *Math Essentials* on www.mathinee.com present important concepts in a clear and concise manner and provide tips on their application. The site also hosts over 400 original problems with full solutions for various levels. Selectors enable the user to sort essentials and problems by test or contest targeted as well as by topic and by the earliest grade level they can be used for.

Online problem solving seminars are available at www.goodsofthemind.com. If you found this booklet useful, you will enjoy the live problem solving seminars.

For supplementary assessment material, look up our problem books in test format. The "Practice Tests in Math Kangaroo Style" are fun to use and have a well organized workflow.

Made in the USA
Middletown, DE
01 February 2016